120
Questions
PART
8
2022

Vibration Analysis Certification Exam Preparation Package

Certified Vibration Analyst Category I

TWO PRACTICE EXAMS

Ali M. Al-Shurafa

Vibration Analysis Certification Exam Preparation Package Certified Vibration Analyst Category I
TWO PRACTICE EXAMS CAT I PREP I SERIES PRACTICE TESTS FOR CVA (PART 8)
ISBN: 978-1-64415-007-8

Disclaimers

Although every precaution has been taken to verify the accuracy of the information contained herein, the author and publisher assume no responsibility for any errors. No liability is assumed for damages that may result from the use of this package. The question banks in this booklet are related to the standard Body of Knowledge according to ISO standard 18436-2 Third Edition 2019.

Title	Vibration Analysis Certification Exam Preparation Package Certified Vibration Analyst Category I: TWO PRACTICE EXAMS	**Author**	Ali M. Al-Shurafa
Subtitle	CAT I PREP I PART 8	**Series**	CAT I PREP I SERIES PRACTICE TESTS FOR CVA
ISBN-10	1-64415-007-7	**ISBN-13**	978-1-64415-007-8
Publisher	Prep Certify	**Ordering**	www.prepcertify.com

Notes

TABLE OF CONTENTS

FORWARD

The International Standardization Organization (ISO) produced the first Body of Knowledge (BoK) for vibration certification in 2003. My relationship with the standard began with ISO subcommittee work, while the standard had draft status. Earlier, a few organizations provided certification in vibration. In fact, I became certified before the ISO, internationally recognized criteria existed.

ISO 184386 18436-2 created four certification levels for those who work in vibration. For those who have not studied vibration—this includes many engineers—related areas formally, such as at a college, these certifications provide a valuable measure of their competence. Certification requirements include training, experience, and exams. Certification exams related to the BoK for each of the four categories of vibration certification.

When I taught a training course pertaining to ISO vibration certification, I would have sample questions and a sample exam. The classes had time to cover the materials, but the students wanted more questions. Developing relevant questions, nay good questions, is a time-consuming challenge. Ali M. Al-Shurafa has risen to this task with a series of eight books that cover the Category 1 BoK used for the certification exams.

Mr. Al-Shurafa divided the series into the eight parts given below:

1. Principles of Vibration
2. Data Acquisition
3. Signal Processing
4. Condition Monitoring
5. Fault Analysis and Correction
6. Rotating Equipment
7. Acceptance Testing
8. Two Practice Exams

As in other specialized fields, part of the difficulty in starting lies in the terminology and basic concepts. This series has questions that aid in understanding. One nominally simple area is the units used in vibration, where it is common to see a variety of units used for the same thing. Many mistakes and misunderstandings arise because of units. Working the problems in the series gives an appreciation for the various meanings of the units used in vibration. Proper use of units is fundamental to progress in the one's career in vibration, and the series involves this thoroughly.

Part 1, Principles of Vibration, has

questions that require the basic understanding of amplitude, frequency, and natural frequencies. The questions have plots from the time and frequency domains. Displacement, velocity, and acceleration measures of amplitude are tested. Metric and U.S. Customary Units are both used. As with other elements of the series, a formula sheet is included, as during an actual exam.

One must collect some data before doing any vibration analysis. **Part 2,** Data Acquisition, treats this area. Vibration analysis requires accurate data. To this aim, questions address selection and proper use of transducers and instrumentation. A critical skill set for those starting out involves the recognition of bad data, which this part of the series covers; It's frustrating to analyze have data returned that doesn't matter represent the condition of the asset.

Part 3, Signal Processing, involves a highly technical subject but covers this well for the first category. This book will help the reader with time domain and frequency domain concepts. It covers the fundamental concepts of FFTs, which are an essential part of vibration work.

Part 4, Condition Monitoring, deals with an ever expanding topic. In relation to vibration, this part for Category 1 focuses on fault recognition. People beginning their certification may not have great expertise in this, but fault recognition provides great value in a condition monitoring program. Grasping the concepts dealt with in this element of the series can provide great value on the job.

While Part 4 concentrates on fault recognition, **Part 5,** Fault Analysis and Correction, explores specific types of faults and the data they can generate. When an anomaly occurs, one must determine the fault condition to address immediate response and any needed corrective action with its urgency. The problems here will help the student develop the required skills.

Many vibration activities concern rotating equipment; for some, rotating equipment vibration comprises the bulk of their jobs. **Part 6,** Rotating Equipment, has problems on a variety of rotating equipment and components, whose understanding is required to diagnose problems in rotating machinery. The variety of equipment the problems address is impressive; the broad range of equipment should suit a variety of individuals from various industries. Working the problems provides a good understanding of basic machinery.

One may find themselves at a vendor's shop with a new machine or in the field with repaired equipment; someone will have to answer whether the equipment meets acceptable criteria. What types of data are needed? What test procedures should be used? **Part 7,** Acceptance Testing, has problems addressing these. The 79 questions in this part are the most I have seen on the topic in any one source.

For those with more training and experience, they may want to start reading at **Part 8,** Two Practice Exams. If one has difficulty in general or specific areas on these exams, the other parts should be consulted. Master the earlier booklets in the series, and the sample exams will provide no obstacle.

Ali M. Al-Shurafa has offered up an extensive set of problems in this series covering the BoK for Category 1 vibration certification to ISO 18436-2. The problems in themselves do not suffice to pass the certification exam, training and study are required to understand these problems. Mastery of these questions, will demonstrate the knowledge required to pass the certification exam. Passing the certification exam after success with these problems should be expected.

I appreciate Mr. Al-Shurafa's efforts

to produce these valuable problems. This will help the next generation of vibration analysts and those just beginning in the field.

William C. Foiles

Sr. Vibration and Rotating Equipment Engineer
Feb 28, 2022

Notes

PREPARATION FOR YOUR CERTIFICATION

The expectations from a Certified Category I Analyst are defined in the ISO 18436-2 standard for both the eligibility for the credential and for the deliverables when the analyst is hired on this category. The compliance in real life is not 100%, however. In the following pages, we'll go through important points to help you understand the certification process with some tips.

To obtain Cat I Analyst credential, the candidate is required to:

1. Have a relevant experience
2. Attend a formal training
3. Pass a written exam

For more details for each of the three requirements, refer to Figure 1.

With respect to the deliverables, the overall expectations from Certified Category I Analyst include being able to perform pre-defined, simple single channel, machinery vibration condition monitoring activities. All activities performed by a Cat I shall be performed under direction in accordance with established procedures.

Figure 1 Certification Requirements for Cat I per ISO 18436-2

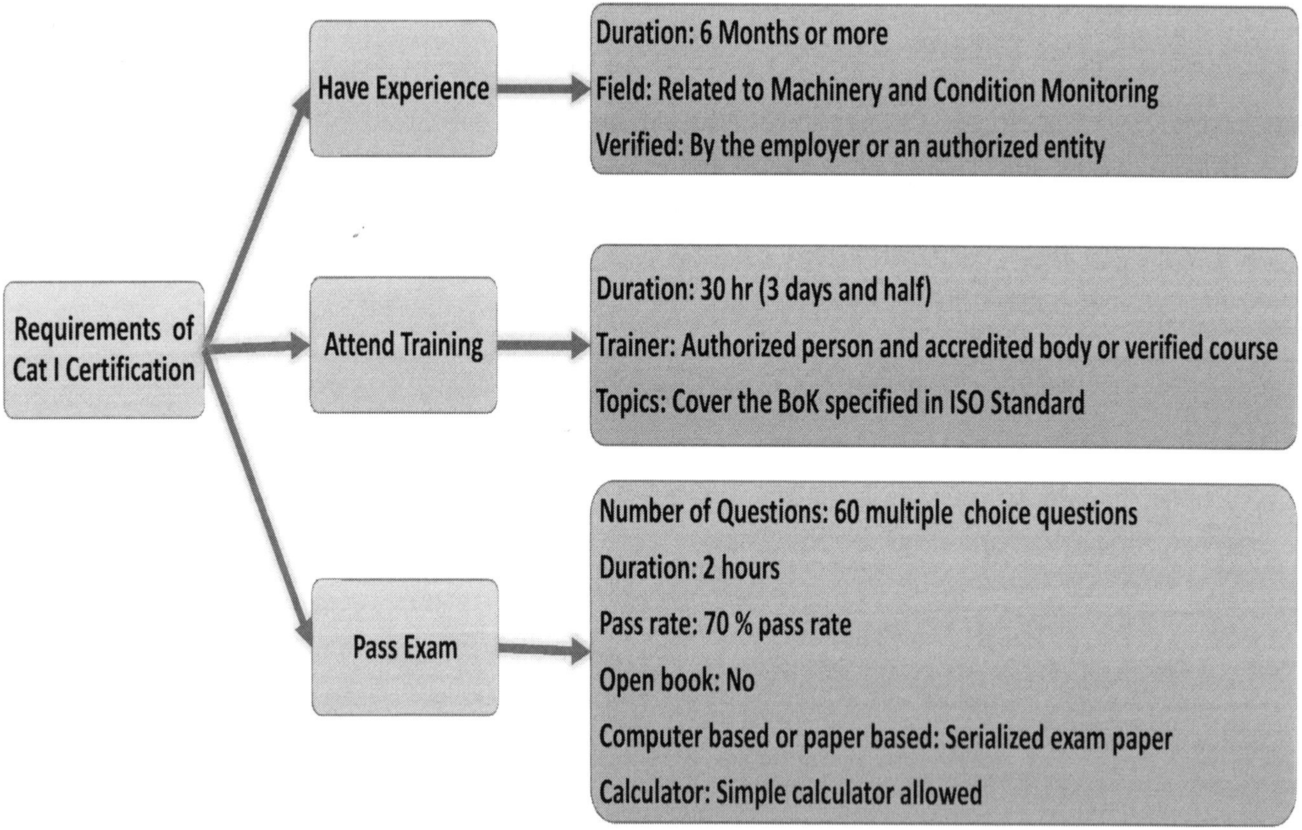

Cat I Analysts are expected to be able to do these specific tasks in Table 1.

#	Table 1- Cat I Tasks
1	Explain the basic principles of mechanical vibration and recognize the different units of measurement
2	Collect reliable data from the machines ensuring appropriate standards of repeatability
3	Identify errors in collected data
4	Retrieve pre-defined measurement settings for data collector or other vibration analysis equipment and transfer data from an analysis system to a computer-based system
5	Compare and evaluate overall or single-value vibration measurements against pre-established alarms
6	Identify deviations from the norm for vibration signals based on single-value or trends
7	Report on visual observations of equipment condition

However, Cat I Analysts are not responsible to do these specific tasks in Table 2.

#	Table 2- Tasks not expected from Cat I
1	Choosing Sensor's type
2	Choosing test method or technique
3	Choosing type of analysis or diagnosis to be conducted
4	Advising on further actions (after the vibration assessment on an asset is done)

To achieve the competency required for the above tasks, the Body of Knowledge for Cat I is provided in the standard. It contains 8 main topics with 25 sub-topics that are included in the training and covered in the exam. The weight of each main topic varies with more than 50% of the total training hours and exam questions dedicated to Principles of vibration, Data acquisition and Equipment knowledge.

Table 3 shows the weighted distribution estimates of the main topics to be spent during the formal training and the number of questions expected in the exam.

The candidates are expected to be familiar with a list of ISO standards which are listed in Table 4.

Table 3 Distribution of topics durign training and number expected exam questions

Book Subject	Distribution Estimates	
	Training Hours	Exam Questions
1. Principles of vibration	6	12
2. Data acquisition	6	12
3. Signal processing	2	4
4. Condition monitoring	2	4
5. Fault analysis	4	8
6. Corrective action	2	4
7. Equipment knowledge	6	12
8. Acceptance testing	2	4
All subjects for Category I	30	60

NOTE: Contents and questions may overlap.

Table 4 ISO standards mentioned as part of the Body of Knowledge for Cat I

#	Code	Title
1	ISO 7919-1	Mechanical vibration of non-reciprocating machines — Measurements on rotating shafts and evaluation criteria — Part 1: General guidelines
2	ISO 10816-1	Mechanical vibration — Evaluation of machine vibration by measurements on non-rotating parts — Part 1: General guidelines
3	ISO 13372	Condition monitoring and diagnostics of machines — Vocabulary
4	ISO 13373-1	Condition monitoring and diagnostics of machines — Vibration condition monitoring — Part 1: General procedures
5	ISO 14694	Industrial fans — Specifications for balance quality and vibration levels
6	ISO 17359	Condition monitoring and diagnostics of machines — General guidelines

For more details on the main and sub-topics, refer to the next section of this book "CAT I PART 8 BODY OF KNOWLEDGE".

To get the most of the two practice exams, please consider the following tips:

1. Detach the empty answer sheet.
2. Complete the exam.
3. Grade your answers compared to the provided answer key.
4. Identify the missed questions and the code on the top of each question.
5. Use Table 5 to link recognized the topis with incorrect answers.
6. Study these topics from the training materials.
7. If you need more practice on certain topic, refer to the relevant part of the series.
8. Retake the exam using the second empty answer sheet.

There are more exam taking trips in Part 1 that you may want to consider.

Good luck,

Ali M. Al-Shurafa

Notes

CAT I PART 8 BODY OF KNOWLEDGE

Table 5 Body of Knowledge for Cat I as specified by ISO 18436-2.

1	Principles of vibration	6 Hours 12 Questions
1.01	Basic motion	Recognize vibration and understand the origin of the sine wave.
1.02	Period, frequency	Recognize the following features of a vibration signal: time axis, period, frequency. Use of hertz or cycles per minute
1.03	Amplitude (peak, peak-to-peak, r.m.s.)	Recognize the following features of a vibration signal: amplitude, peak, peak-to-peak, r.m.s.
1.04	Parameters (displacement, velocity, acceleration)	Recognize the following parameters: displacement, velocity, and acceleration
1.05	Units, unit conversions	Recognize that units conversion is possible
1.06	Time and frequency domains	Be aware of time and frequency domain
1.09	Natural frequency, resonance, critical speeds	Be aware resonance exists, and its effect on vibration

2	Data acquisition	6 Hours 12 Questions
2.01	Instrumentation	Recognizing single- channel hand-held route-based and on- line measurement and monitoring systems
2.03	Transducers	Recognize displacement, velocity and acceleration transducers. Be aware of powered and non- powered types
2.04	Sensor mounting, mounted natural frequency	Recognize broad effects of mounting on the frequency response, e.g. stud, magnet or probe
2.09	Test procedures	Follow pre-set data acquisition procedures for on-line or route- based systems. Recognize measurement points for common machine types. Recognize some poor data and alarm conditions. Be aware calibration is a requirement
2.11	Computer database upload/down- load	Be aware of basic functions of host and data collector
2.12	Recognition of poor data	Recognize simple fault conditions, e.g. ski-ramp, no signal, cable fault
3	Signal processing	2 Hours 4 Questions
3.05	FFT application	Be aware of the term FFT and recognize the following basic FFT terminology, e.g. number of lines, Fmax and time to sample

4	Condition monitoring	2 Hours 4 Questions
4.08	Fault condition recognition	Recognizing basic pre-set fault conditions, e.g. unbalance, looseness, misalignment, bearing noise and damage
5	Fault analysis	4 Hours 8 Questions
5.18	General fault recognition	Recognize fault frequencies for pre- set FFT and simple time waveforms for unbalance, looseness, misalignment, bearing noise and damage. Also recognize the terms: resonance and phase
6	Corrective action	2 Hours 4 Questions
6.07	Basic maintenance action	Be aware of simple maintenance actions to rectify or reduce faults e.g. lubrication, alignment
7	Equipment knowledge	6 Hours 12 Questions
7.01	Electric motors, generators and drives	Recognize AC induction motor, and basic faults, e.g. bearing noise and damage, balance, looseness and misalignment
7.02	Pumps, fans	Recognize basic pump and fan combination , and basic faults, e.g. bearing noise and damage, balance, looseness and misalignment
7.04	Compressors	Recognize examples of centrifugal and screw compressors

7.06	Rolling mills, paper machines, other process equipment	Recognize examples of centrifugal and screw compressors of these machines
7.07	Machine tools	Recognize examples of these machines
7.08	Structures, piping	Recognize the term: resonance
7.09	Gearboxes	Recognize basic examples of simple gearboxes
8	Acceptance testing	2 Hours 4 Questions
8.1	Test procedure	Be able to apply basic pre-set methods, and be aware of access and safety requirements

CAT I BODY OF KNOWLEDGE

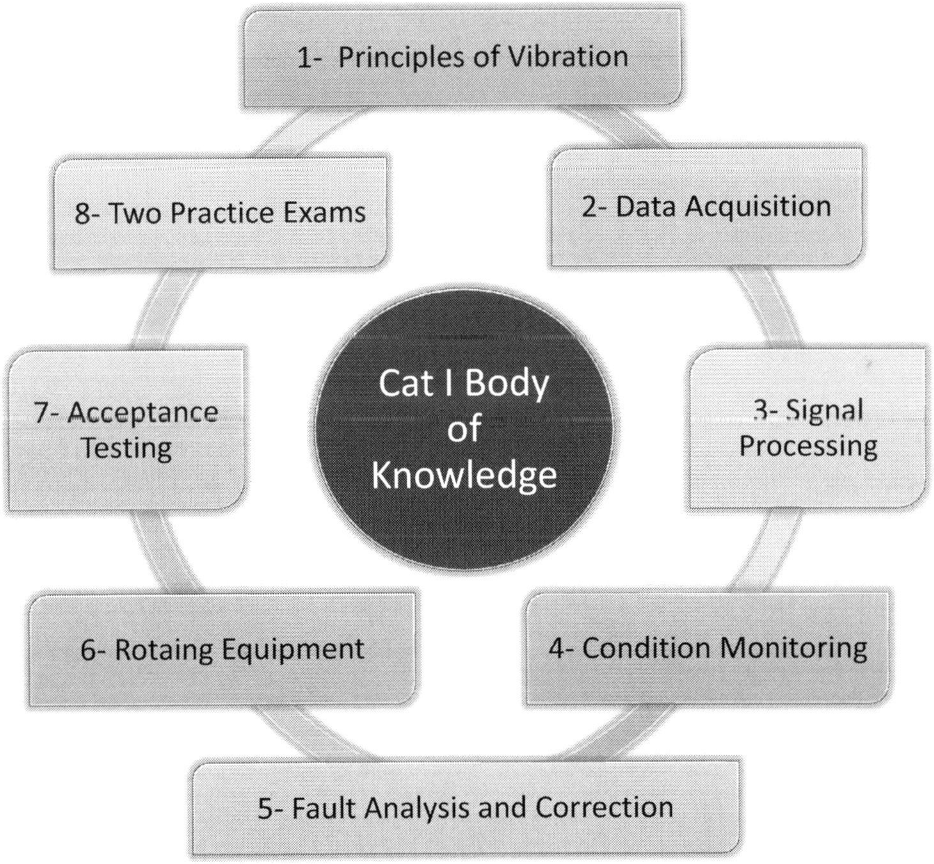

FORMULA SHEET FOR CAT I PREP I PART 8

Constants:

1 Hz = 1 cps = 60 cpm = 60 rpm

1 minute = 60 seconds = 60,000 milliseconds

$\pi \approx 3.14$ 1 rad/sec ≈ 9.55 cpm $g \approx 386.1\ \text{in/sec}^2$

Equations:

$$\text{Order nx} = \frac{F}{\text{Speed}} \qquad F = \frac{\text{Cycles}}{T} \qquad P = \frac{1}{f}$$

Order = unitless (ratio) F = frequency of vibration (cpm)

Speed = Shaft rotational frequency (rpm)

T = time covered (sec) Cycles = count of cycles captured over time T

P = period (sec) for a single complete cycle

f = frequency (Hz)

Simplified General Motion Equation

$$m * a + c * v + k * d = f\ (t)$$

m = mass (lbm),
a = acceleration (g)
f(t) = force function of time (lbf)
c = damping (lb.sec/in)
v = velocity (ips)
k = stiffness (lb/in)
d = displacement (mil)

Units shown above are for reference only and should not be used for the general motion equation.

Amplitude conversions for single freuency in Hz

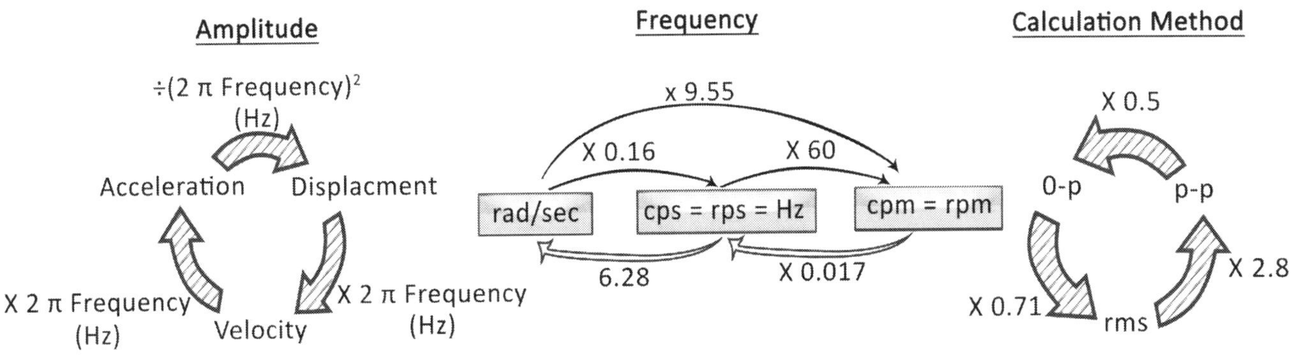

Constants:

1 minute = 60 seconds = 60,000 milliseconds

1 Hz = 1 cps = 60 cpm

Equations:

p= Period (sec) $p = \frac{1}{f}$ f= frequency (Hz)

Resolutior $2 * WF \frac{F_{MAX}}{LINES}$

DAT = Lin(

Window	Factor
Hanning	1.5
Flat tap	3.8
Uniform	1

Res = Resolution (Hz)

F_{max} = Maximum Frequency (Hz)

WF = Window Factor

DAT = Data acquisition time (sec)

Lines = Count of resolution lines = LOR

TYPICAL FAULT FREQUENCY ORDERS

Frequency Order	Possible Fault
Almost zero	False signal. Sudden large movement. Ski slope.
0.3 - 0.7 X	Cage defect of a rolling element bearing
1X	Many faults. Unbalance. Misalignment
2X	Misalignment.
NX (3 - 8)	Misalignment. Vane or blade pass frequency Looseness.
NX (20-100) N: Integer	Many faults. Vane or blade pass frequency. Rotor bar pass, Gear mesh frequency
NX (3.01-30.1) N: non-integer	Inner race, outer race, ball pass frequency and background excitation.
Fault	Possible Order or other symptoms
Rotor unbalance	1X horizontal and vertical similar amplitudes but can be higher in one direction.
Shaft misalignment	Usually 2X or 3X but can be 1X, radial and possibly axial

Rolling element bearing defect	Non-integer order from 0.25 X - 8.9 X. Usually localized on the bearing of defect. Noisy waveform and spectrum.
Process upset on pump, fan and compressors	Vane or blade pass frequency. Many times with noisy waveform and spectrum.
Slip for an induction motor	Synch speed - Running speed
Synchronous Speed for motors and generators	SN = (2 x LF) / Poles

120
Questions
PART
8
2022

EXAM A

TWO PRACTICE EXAMS

Ali M. Al-Shurafa

Practice Exam A

Question 001

1.01

Which of the following is a sine signal?

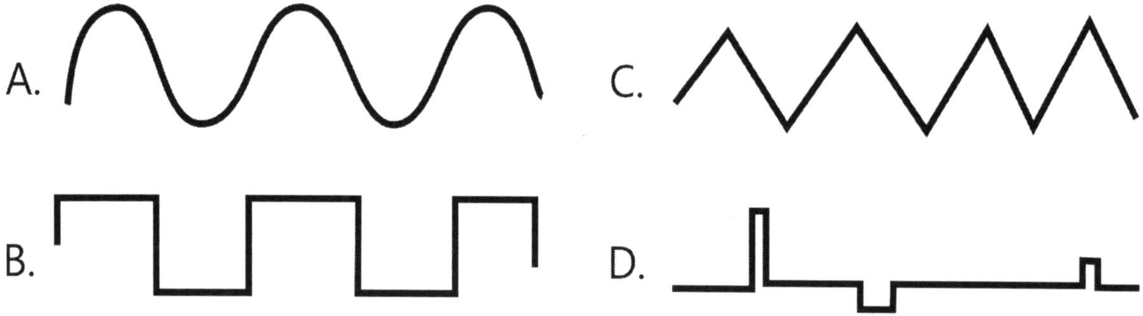

Note

1.01 1.02 1.04

Practice Exam A
Question

As a mechanical fault develops (gets worse) in a rotating machine, what is the most important change an analyst would see in the vibration plots?

A. An increase in the overall amplitude.

B. An increase in F_{max}.

C. An increase in the phase of the 1X running speed.

D. A reduction in the resolution.

Note

Practice Exam A

Question 003 1.01 1.04

Consider typical industrial vibration applications for machines running at speeds between 20 - 60 Hz. Which measurement is the best to represent each of the following?

	Fatigue of Mechanical Components	Exciting Forces
A.	Displacement	Displacement
B.	Velocity	Velocity
C.	Displacement	Acceleration
D.	Velocity	Acceleration

Note

1.02 1.06

Practice Exam A
Question

Which waveform has a high amplitude and a high frequency? Assume all plots share the same scales.

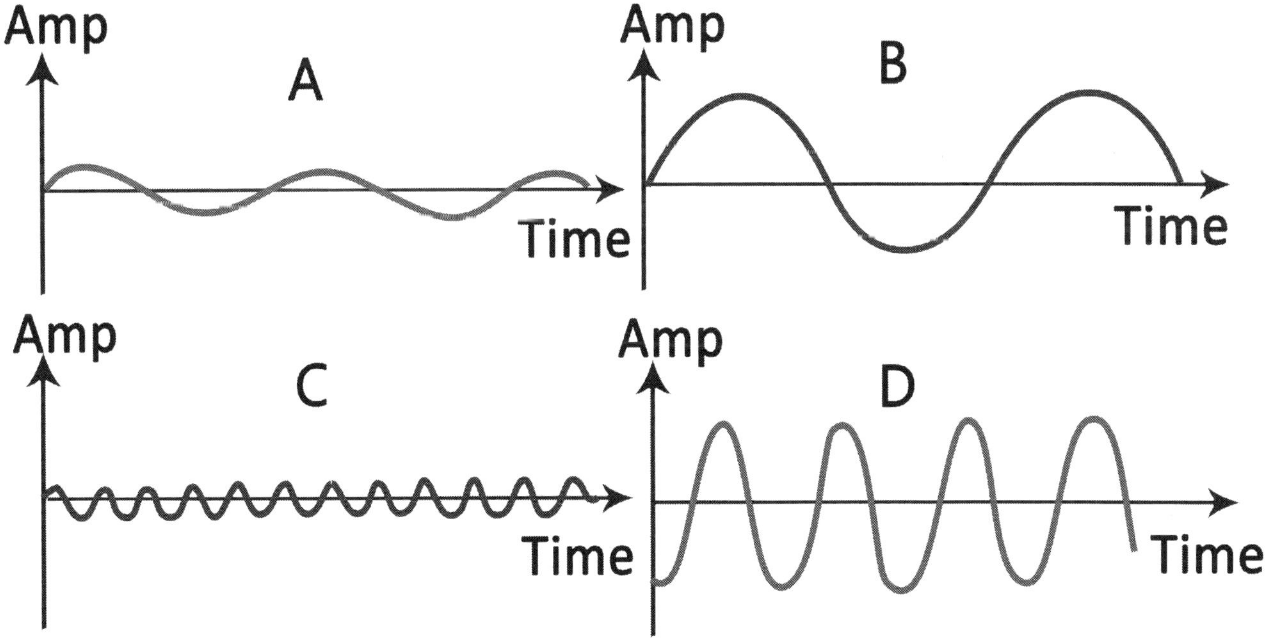

Practice Exam A
Question 005

1.02 | 1.06 | 3.05

Study the provided vibration plot. What is the data acquisition time?

A. 0.1 sec

B. 17 sec

C. 100 sec

D. 0.017 sec

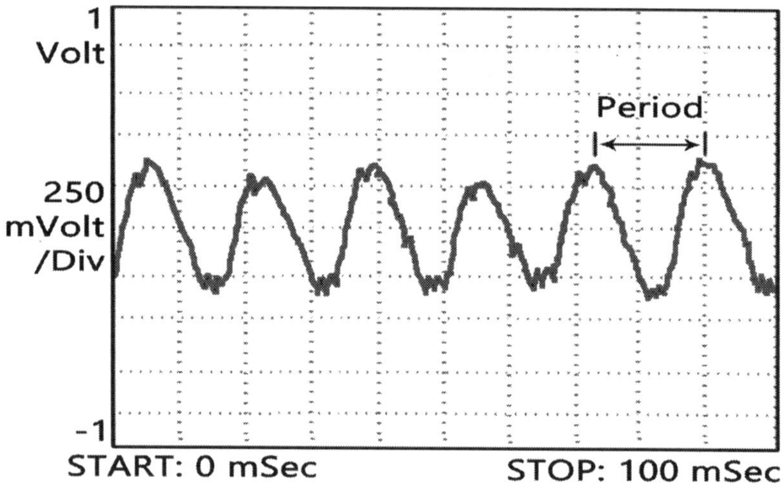

Note

1.02 1.05

Practice Exam A
Question

Assume a shaft vibrates an amount of 100 micrometers up and down 100 times in a minute. What is the frequency of this vibration?

A. 100 Hz

B. 100 cpm

C. 1 micrometer / minute

D. 100 microns p-p

Note

Practice Exam A
Question 007

1.03

Study the vibration plot. What is the overall amplitude in peak? Sometimes, this value is called true peak. Expect some rounding.

A. -0.39 inch/sec

B. 0.22 in/s

C. 0.38 ips

D. 0.6 inch per sec

Note

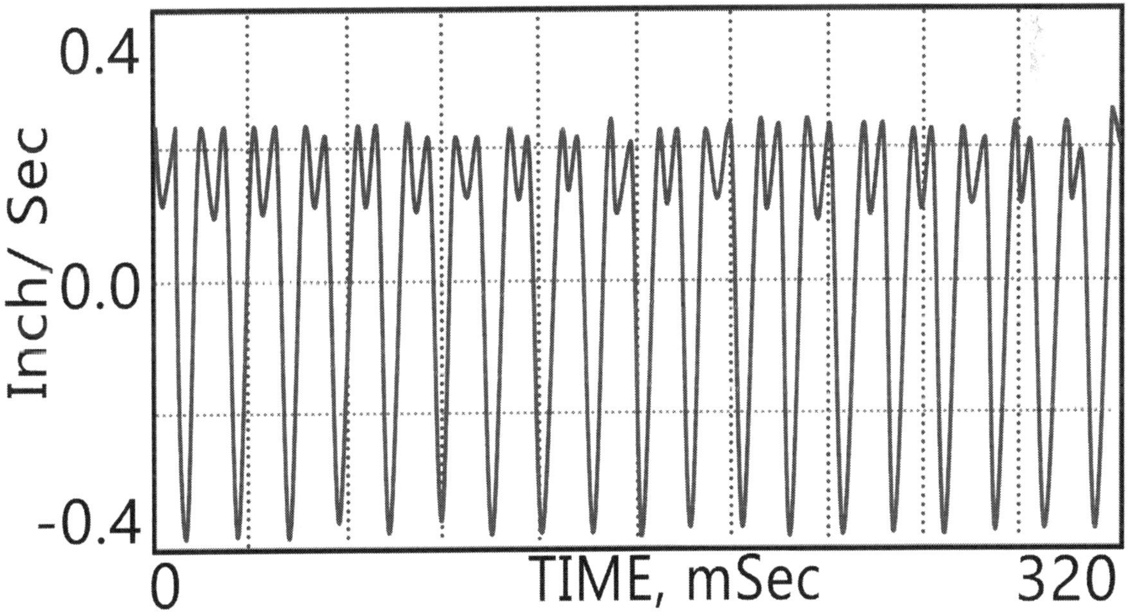

Note

Practice Exam A
Question 008

1.02 1.05 1.06

Study the next vibration plot. What is the frequency in CPM and Order? Expect some rounding.

	CPM	Order
A.	2.5	3.8X
B.	360	2.9X
C.	3.7	1X
D.	16220	1X

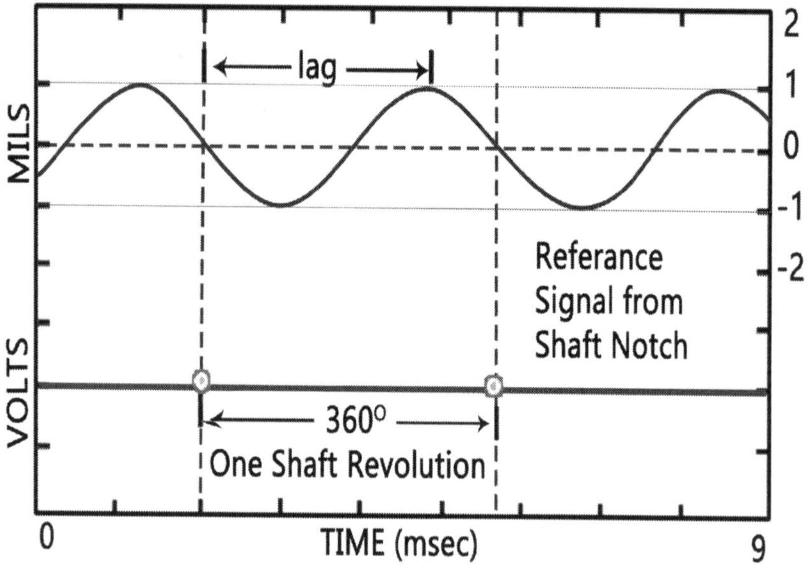

1.03 1.05

Practice Exam A
Question

Study the next plot. What is the amplitude of 1X?

A. 14 mm/s 0-p

B. 10 mm/s rms

C. 20 mm/s rms

D. 20 mm/s 0-p

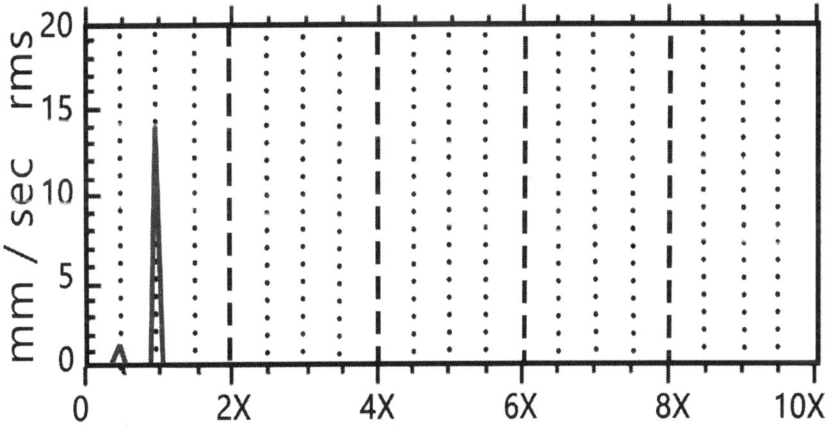

Note

Practice Exam A

Question 010

1.02 1.06

Refer to the next vibration plot for a small fan. What is the frequency expressed in Hz for the component of the highest amplitude? The fan's speed is 3598 revolutions per minute.

A. 30

B. 60

C. 120

D. 0.5 X

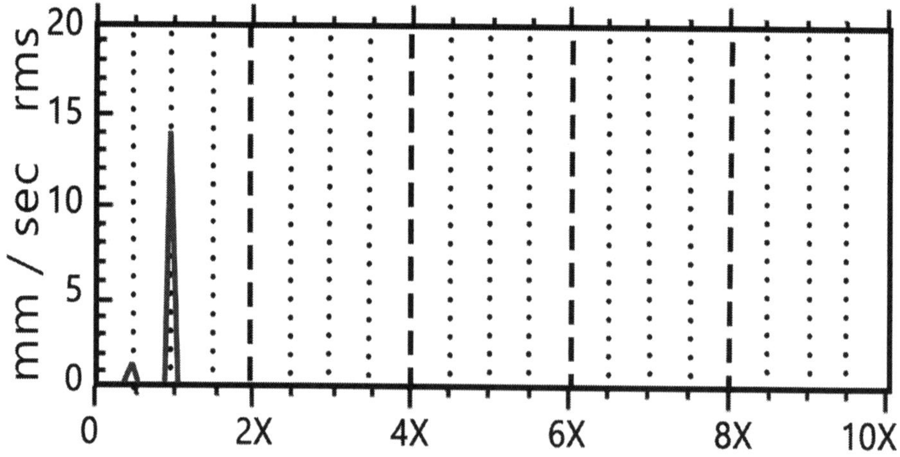

Note

1.05 5.18

Practice Exam A

Question

What is a typical application of "phase" in practical vibration analysis?

A. To distinguish between similar machinery faults (e.g. misalignment and unbalance).

B. To perform rotor balancing (in shops and on sites).

C. To analyze transient data (i.e. startup and shutdown).

D. All of the above.

Note

Practice Exam A

Question 012

2.01 | 2.11

In a plant's vibration monitoring program, why is vibration data transferred from the data collector to a computer?

A. To build the vibration history of the machines (database)

B. To prevent equipment failures

C. To perform more detailed analysis

D. A and C

Note

2.01 2.09 8.1

Practice Exam A
Question

For a proper vibration evaluation of rotating machines like fans, how many points are typically collected?

A. 3 horizontal and 3 vertical readings

B. 3 points on the driver and 3 points on the driven machine

C. 3 readings per bearing

D. A reading per 1000 rpm. Example: 3 readings if the machine's speed is 3600 rpm.

Note

Practice Exam A

Question 014

2.01	2.03	2.09	

In the industry, what is another technical term used for a "shaft vibration measuring system"?

A. Non-contact or relative probe system

B. Displacement or proximity probe system

C. Eddy current probe system

D. All of the above

Note

2.03

Practice Exam A

Question

Which type of vibration sensors typically needs an external electric power source to function properly?

A. Relative probe

B. Accelerometer

C. Velocity transducer

D. A and B

Note

Practice Exam A
Question 016

2.01	2.09	8.1	

An 8-pole induction motor (895 rpm) drives a huge fan (2000 HP) through a simple gearbox. The shafts (of the three machines) are mounted on journal bearings. The power line is 60 Hz.

Based on this limited information, what is the most appropriate vibration measure to be used for the motor?

A. Shaft axial position

B. Shaft displacement

C. Shaft velocity

D. Bearing housing acceleration

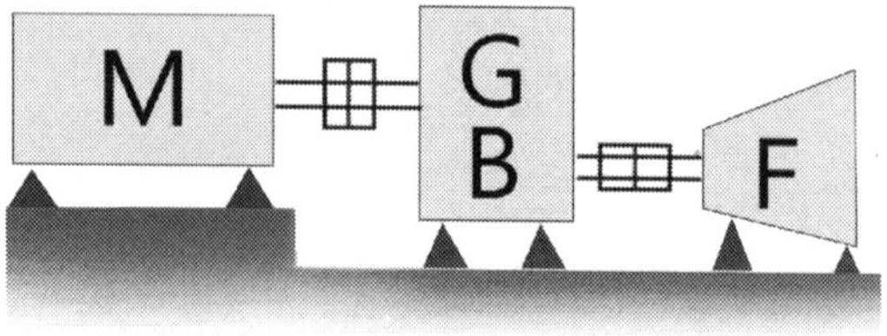

2.01 2.09

Practice Exam A

Question

In some vibration programs, measurements are given certain codes like 1, 2, 3 and 4. Following are example measurements from a pump: P1H, P1V, P1A, P2H, P2V and P1A. What do these numbers most likely refer to?

A. Usually these numbers refer to the bearings of the machine in a certain sequence. For example outboard bearing is 1 and inboard bearing is 2.

B. The number refers to the criticality of the vibration measurement. Usually, the higher the number, the higher the criticality.

C. This number refers to the number of averages needed on that specific point.

D. This number refers to the number of spectra needed for that specific point.

Practice Exam A

Question 018

2.09 | 7.02

What is another term used for "pump outboard" bearing? Refer to the next illustration.

A. Non-drive end

B. Drive end

C. Coupling side

D. Discharge side

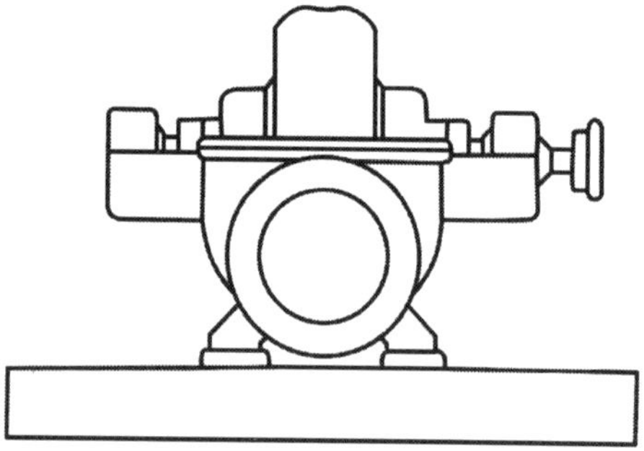

Note

2.09 7.01

Practice Exam A
Question

Refer to the next illustration for a motor-pump unit. What is the best designation for the vibration measurement point marked by the arrow?

A. Motor NDE axial

B. Motor top inline

C. Pump top axial

D. Pump non-drive end velocity

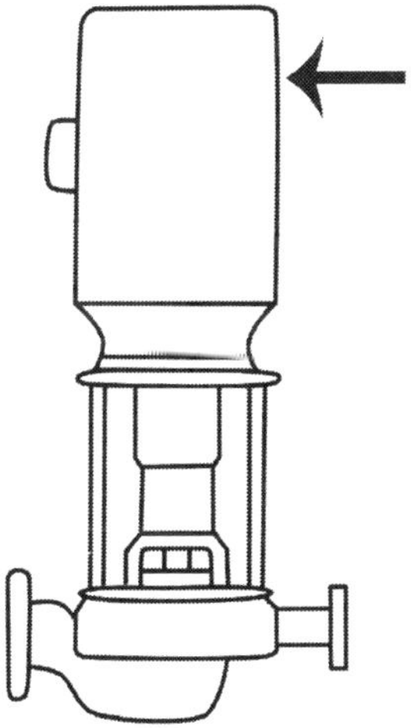

Note

Practice Exam A
Question 020

2.09	7.02	8.01	

Refer to the next illustration for a between bearing centrifugal pump. Which point is referred to as "pump non-drive end axial"? "Pump outboard vertical"?

	Pump non-drive end axial	Pump outboard vertical
A.	1	5
B.	1	6
C.	4	2
D.	2	5

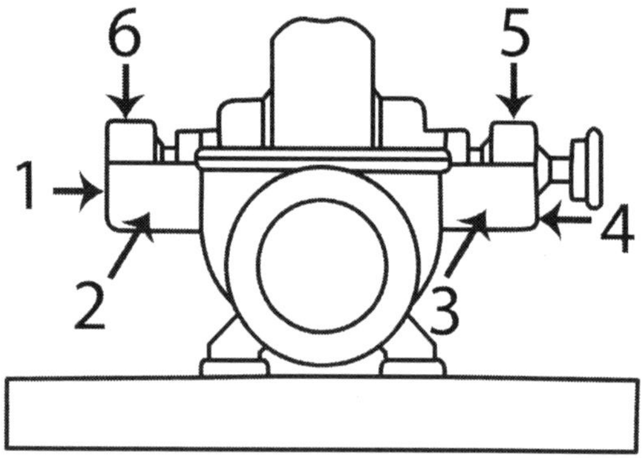

Note

2.01 2.03 2.04

Practice Exam A
Question

An instrument technician reported that Vibration Sensor 15-V-151 is stud mounted and it has a sensitivity of 100 mV/g according to its data sheet. What type of vibration sensing element is this?

A. An accelerometer that produces 100 millivolts for each vibration cycle.

B. An accelerometer that produces one volt for each 10 g's of vibration.

C. A relative probe whose gap between the shaft and probe's tip is g = 100 mils.

D. A speed pickup able to detect 100 teeth in a gear.

Note

Practice Exam A
Question 022

2.01	2.03	2.09	

For an accelerometer used in a periodic vibration data collection program, what is the meaning of: "two-pole magnet"?

A. The number of north and south sides (electromagnetic) in the winding of an electric motor.

B. A standard type of data transfer cables that comprises two signal wires and a third wire for grounding.

C. A mounting device (made of magnetic material) used to hold vibration sensors on bearing housings.

D. A signal processing property configured in the analyzer to collect data between two frequencies.

Question

Consider a classic vibration data collection from a group of machines. How significant is "consistency" when you apply the measurement procedure?

A. It is very essential to establish reliable trends.

B. It is very essential to detect qualitative and quantitative changes in a timely manner.

C. It is very essential in order to maintain the finest resolution (in frequency domain) and reduce data collection time (in time domain). Consistency establishes repeatability.

D. A and B

Practice Exam A

Question 024

2.09	2.12	8.1

While you are at the site collecting vibration data using a portable data collector, the accelerometer dropped from your hand on the ground. What should be your first reaction from the given answers?

A. Go back to the supervisor and ask for a new accelerometer.

B. Send the accelerometer to the vendor for calibration.

C. Test the accelerometer on one of the machines you have just measured.

D. Proceed with the data collection. Drops of a vibration sensor are normal.

Note

2.04 2.12 8.12

Practice Exam A

Question

Consider a small fan with historical vibration amplitudes from 0.08 to 0.10 inch per second. Your current measurement is 0.02 inch per second. Alarm is set at 0.35 ips. What should be your first response to this reading while you are next to the machine? Select from the given answers.

A. Reading is low and excellent. No need to collect more readings.

B. Reading is better than history and within the limit. I'll move to the next point.

C. Reading is too low. I'll take another reading to confirm the value and send a report about it.

D. Reading is too low. I'll go back to the office and replace the sensor to take a new reading.

Practice Exam A

Question 026

2.01 2.04 2.12

Which part of the vibration instrument (used in route measurements) can introduce an error in the reading?

A. Sensing element

B. Sensor's mounting device

C. Cable

D. All of the above

Note

2.09 2.12 3.05

Question

What does a ski slope in the vibration spectrum indicate?

A. It is a high amplitude at the first bin indicating a possible fault in the instrument loop.

B. It is a high vibration amplitude at very low frequency indicating unbalance.

C. It is a high vibration amplitude at 1X indicating unbalance.

D. It contains two closely spaced frequencies acting on the shaft indicating a coupling fault.

Note

Practice Exam A

Question 028

2.09 5.18

Study of the following machinery vibration waveform plots. Which one is the best to demonstrate a possible problem in the data collection process? Assume all plots share the same scales.

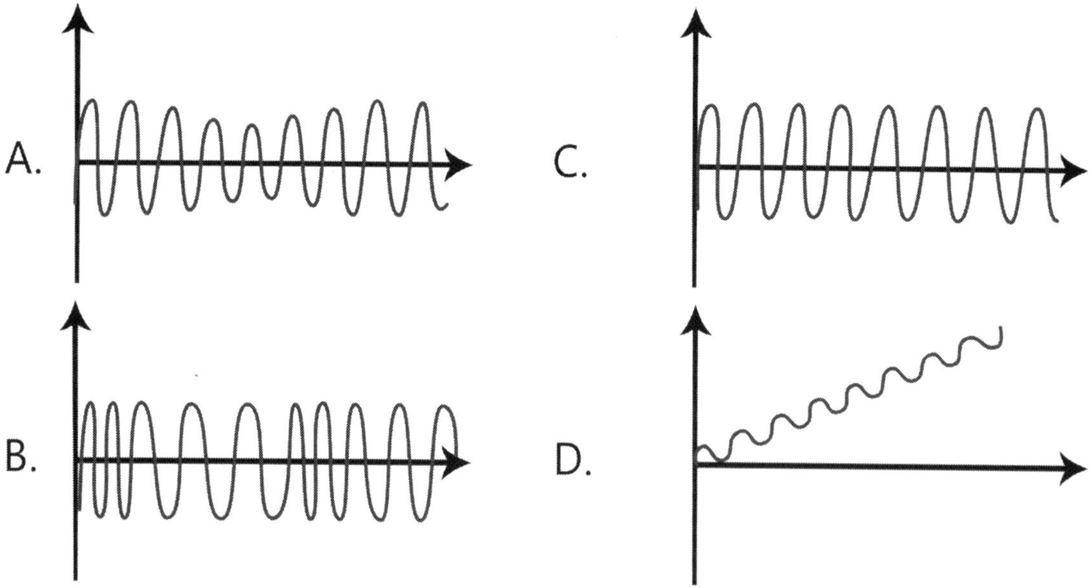

2.09 5.18

Practice Exam A
Question

What is the best description for the below spectrum taken from a pump? Select from the below expressions.

A. High harmonic components

B. High ski slope

C. High sidebands

D. High noise floor

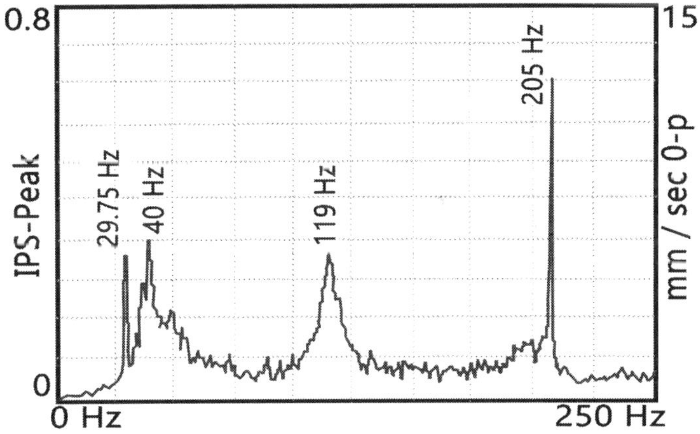

Note

Practice Exam A

Question 030

1.06 | 3.05

What is the mathematical algorithm used by vibration analyzers to create a spectrum?

A. Fast Frequency Technology. It transforms data from time domain to frequency domain.

B. Fast Fourier Transform. It transforms data from a waveform to a digital spectrum.

C. Fast Fourier Transform. It transforms vibration from displacement to velocity.

D. FORTRAN Frequency Transfer. It transforms spectrum from 0-p to rms.

Note

3.05	1.06	5.18

Question

Study the next vibration plot. What is the domain and natural frequency?

	Domain	Natural Frequency
A.	Frequency	Unknown
B.	Velocity	10 X
C.	Frequency	1 X
D.	14.8 mm/sec	1 X

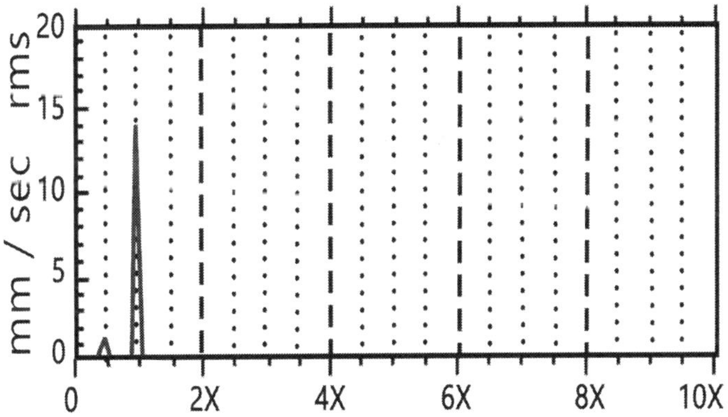

Note

Practice Exam A

Question

1.02	4.08	5.18

In machinery vibration analysis, what is the meaning of a "sub-synchronous" signal and "synchronous" signal? Select the best answer from the given choices.

	Sub-synchronous	Synchronous
A.	Frequency <1X Shaft Speed	Frequency = 1X Shaft Speed
B.	Horizontal Amplitude < Vertical Amplitude	Horizontal Amplitude = Vertical Amplitude
C.	Current Amplitude < Average Amplitude	Current Amplitude = Average Amplitude
D.	Preprocessed Amplitude	Being Processed Amplitude

Note

3.05 4.08 5.18

Which of the following spectra demonstrates a non-synchronous signal for a motor running at 59.5 Hz? Assume linear y-axis scales.

A.

B.
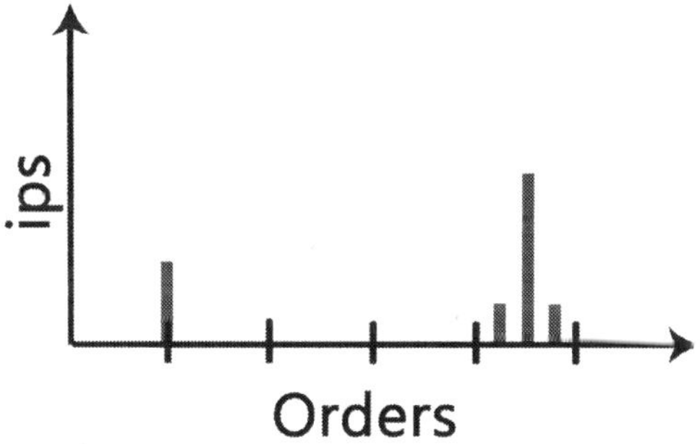

C.
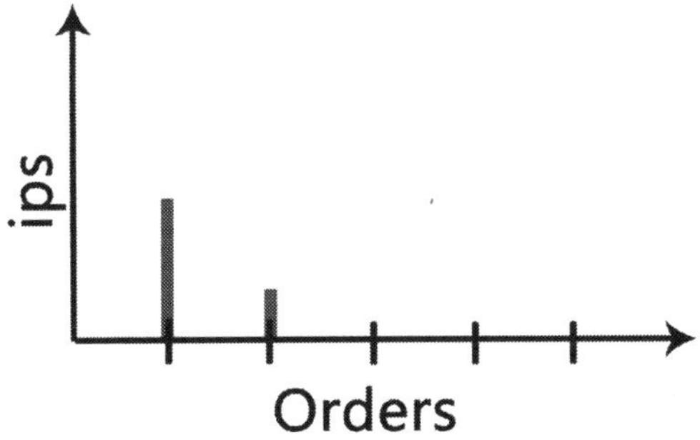

D.

Practice Exam A

Question 034

3.05	5.18	7.01

What is a typical frequency range (F_{min} - F_{max}) used for basic vibration analysis for a medium size electric motor (5-1000 hp)? Select the best answer from the given choices.

A. 0.005 Hz to 0.05 Hz

B. 0.05 Hz to 5 Hz

C. 5 Hz to 1000 Hz

D. 5 kHz to 1000 kHz

Note

3.05

Practice Exam A
Question

For a classic machinery vibration waveform collected by a digital analyzer, what do 1024 samples represent?

A. The total number of data points (x,y) displayed in a spectrum is 1024.

B. The total number of data points (x,y) displayed in a waveform plot is 1024.

C. The resolution of the y-axis (amplitude) in the waveform plot is 1024 points.

D. The total time required to sample the waveform is 1024 mseconds.

Note

Practice Exam A

Question 036

1.02 | 1.05 | 3.05

Study the next waveforms and spectra. Which plots are related? Expect some rounding.

A. 1 and 3

B. 1 and 4

C. 2 and 3

D. 2 and 4

Note

1.

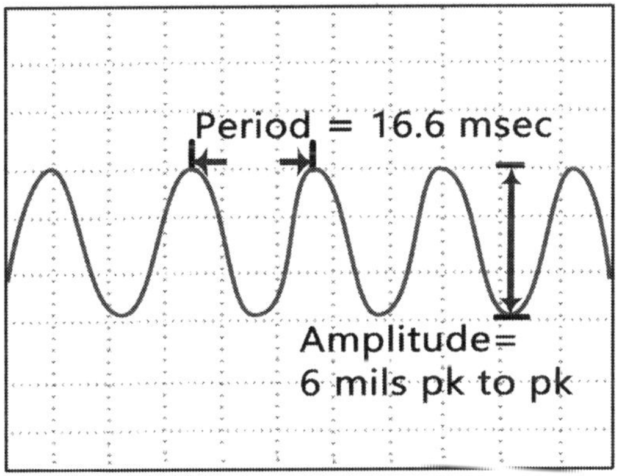
Period = 16.6 msec
Amplitude=
6 mils pk to pk

2.

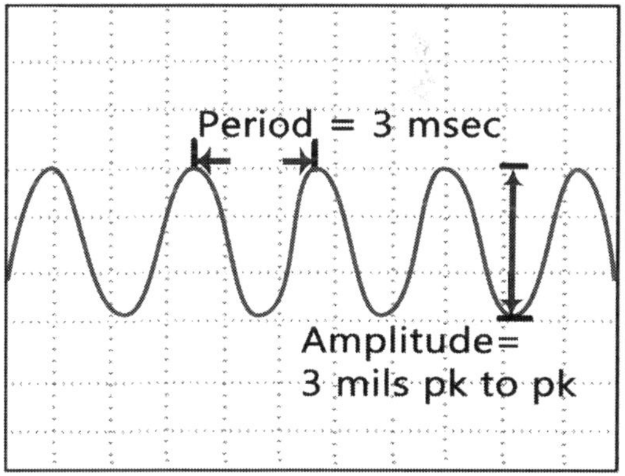
Period = 3 msec
Amplitude=
3 mils pk to pk

3.

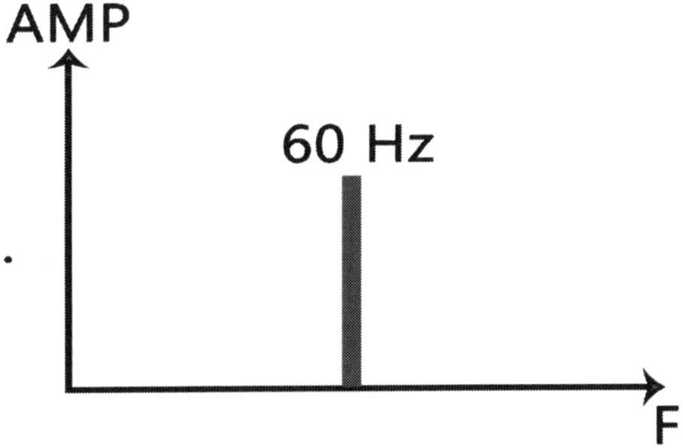
AMP
60 Hz
F

4.

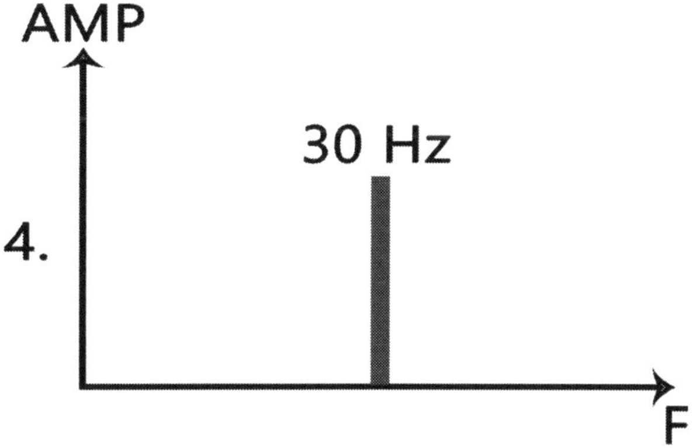
AMP
30 Hz
F

Practice Exam A

Question 037

7.09

What is the gear mesh frequency for the following simple gearbox? Input shaft has 23 teeth and it turns at 1000 rpm while its output shaft has 57 teeth. Hint: Add the numbers to the illustration.

A. 23000 cpm

B. 23000 HZ

C. 57000 cpm

D. 57000 Hz

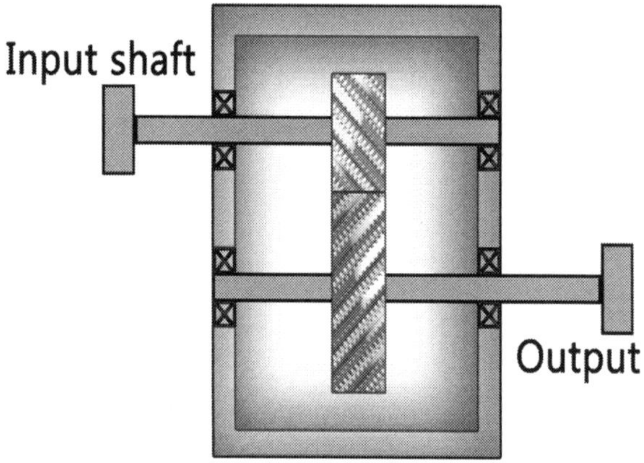

Note

7.09

Practice Exam A
Question

Study the shown unit of a motor driving a compressor through a gearbox. Which side of this gearbox is referred to as output shaft?

A. The low speed shaft

B. The high speed shaft

C. The shaft with the higher power

D. The shaft connected to the motor

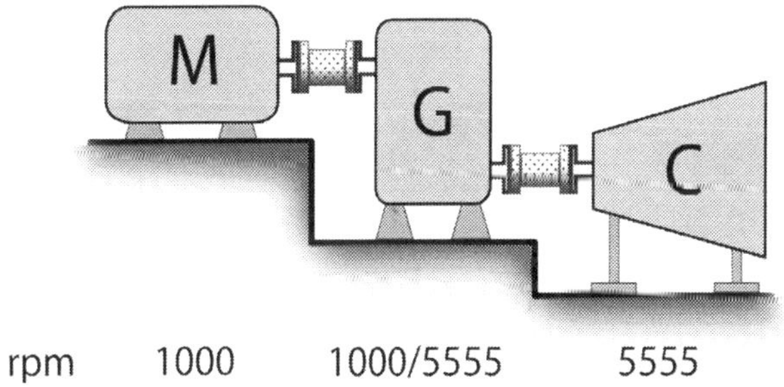

Note

Practice Exam A

Question 039

7.04

If this impeller rotates at 52300 rpm, what is its vane pass fault frequency? This type of impeller is commonly used in plant air compressors. Hint: count the vanes.

A. 15690 Hz

B. 15690 cpm

C. 52300 Hz

D. 52300 cpm

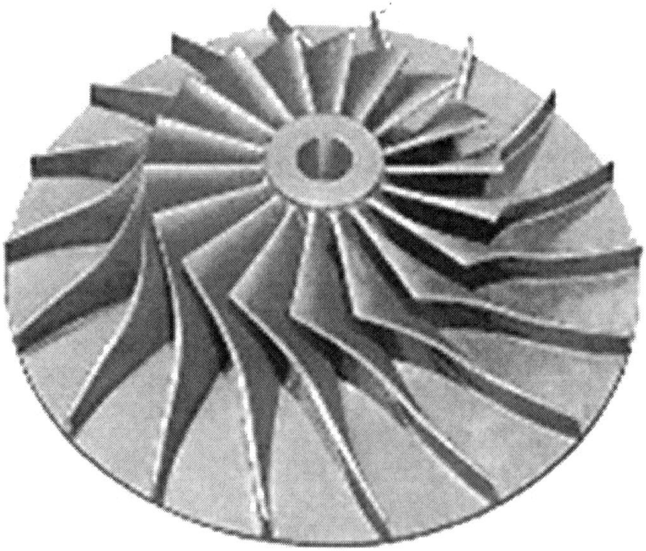

6.07 7.01

Practice Exam A
Question

What is a function of a radial bearing in rotating equipment? Refer to the illustrations. Select the best answer from the given choices.

A. It allows the rotational motion of the shaft and minimizes its translational motions.

B. It holds the parts of the rotor and rotates them.

C. It converts the thermal/hydraulic energy to rotational energy and vice versa.

D. It is transmits the power from the shaft to the bearing housing.

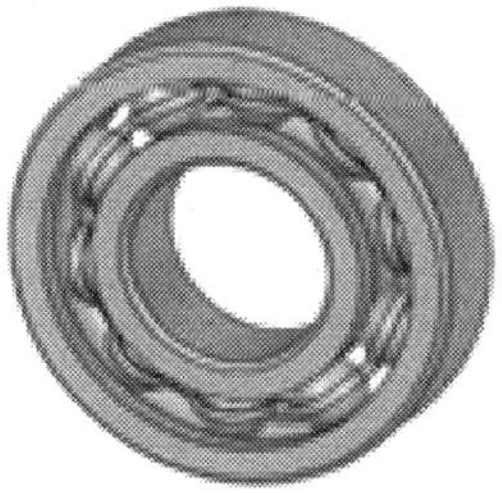

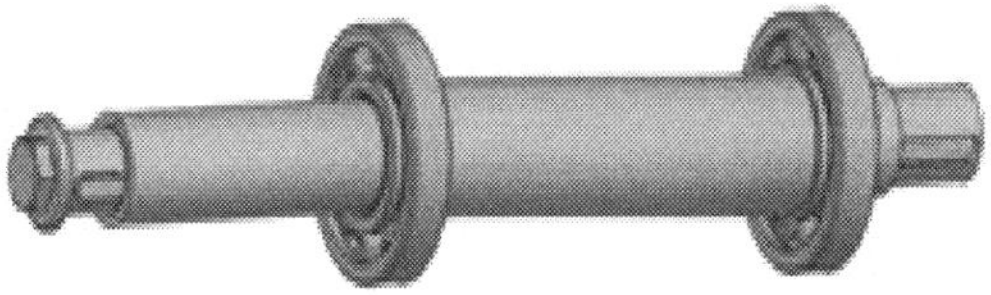

Practice Exam A

Question 041

7.09

Refer to the next illustration for a typical simple gearbox. How many radial bearings are used to carry the shafts?

A. 2 bearings

B. 4 bearings

C. 8 bearings

D. 16 bearings

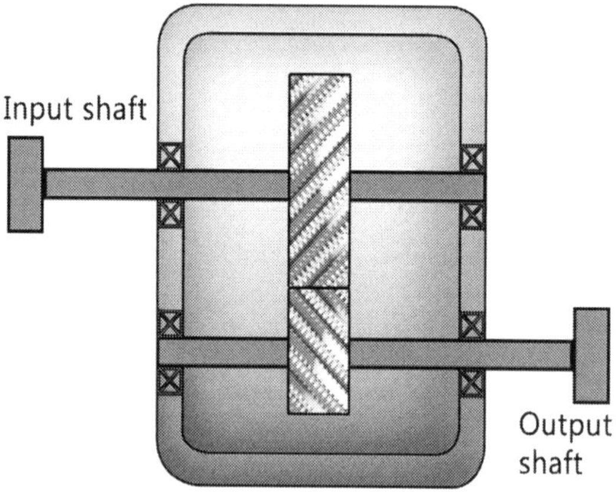

Note

7.04

Question

What are three major parts of a typical centrifugal compressor?

A.	Rotor	Winding	Bearings
B.	Plunger	Inlet valve	Outlet valve
C.	Pressure gauge	Temperature indecator	Flow meter
D.	Shaft	Impeller	Seal

Note

Practice Exam A

Question 043

7.01 | 7.02

What component of rotating equipment is shown in the below illustration? What is its function?

	Part	Function
A.	Rubber coupling	To transmit power
B.	Rolling element bearing	To hold shaft
C.	Rubber gearbox	To store power
D.	Bearing housing	To position the bearing

7.06

Practice Exam A
Question

What is the most common kind of bearings used for rotating equipment in the following industries: printing, paper and rolling mills?

A. Tilting pad bearing

B. Air lubricated bearing

C. Bevel bearing

D. Rolling element bearing

Note

Practice Exam A

Question 045

7.01 | 7.08

What are the main parts of a typical ball bearing? Refer to the illustrations.

A. Upper shell, lower shell, balls and liner

B. Balls, rollers, needles and spheres

C. Driver hub, driven hub, shims and spacer

D. Inner race, outer race, cage and rolling elements

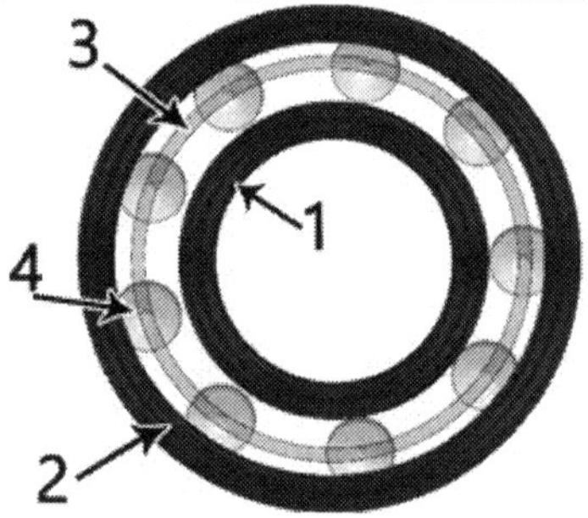

4.08 5.18

Practice Exam A
Question

Which symptom is the best indication to recognize a defective ball bearing from its vibration spectrum? Refer to the next illustration for a ball bearing (10 balls) inside a motor.

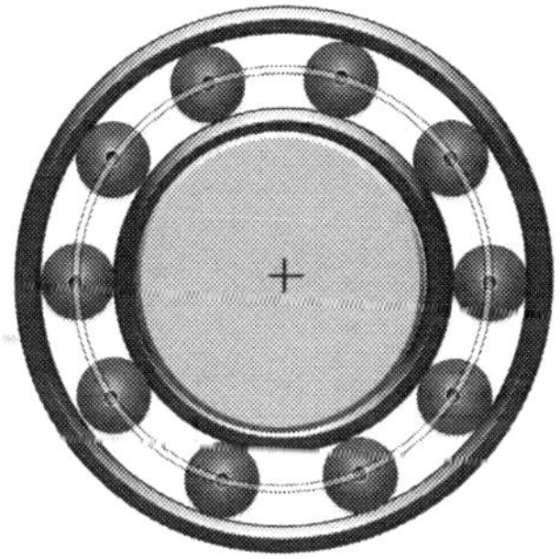

A. Strong presence of non-synchronous components

B. Strong presence of synchronous components

C. Strong presence of 2 line frequency components

D. Strong presence of 10X components (10 balls)

Note

Practice Exam A

Question 047

1.09 5.18

Consider a 20 Hp (15 kW) motor driving a horizontal pump. In practice, which of the following can be a corrective action for structural resonance and looseness?

	Resonance	Looseness
A.	Increase resolution	Add or remove weight
B.	Improve supports	Tighten bolts
C.	Grease bearings	Increase gap voltage
D.	Increase F_{max}	Increase suction pressure

Note

5.18

Practice Exam A
Question

Which plot of the following vibration waveforms is the best to demonstrate a possible machine trip?

A.

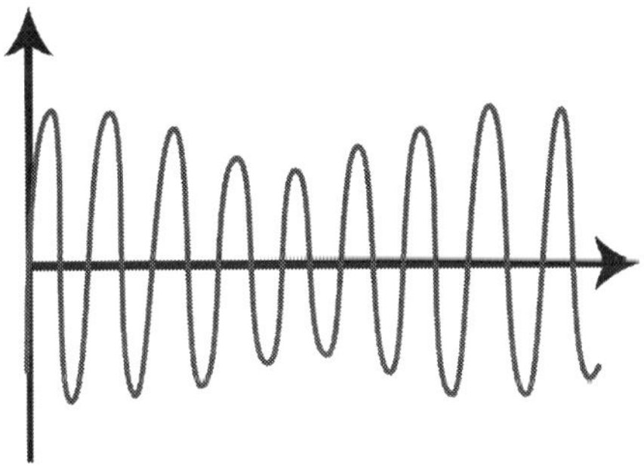

B.

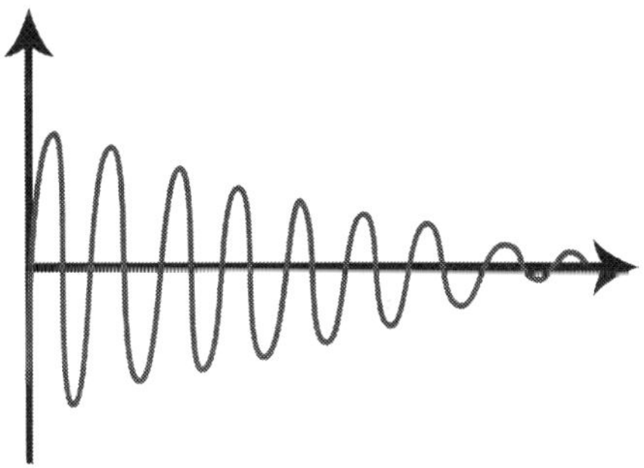

C.

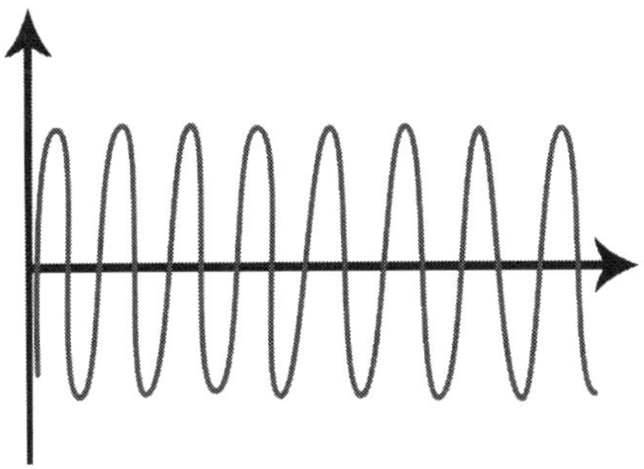

D.

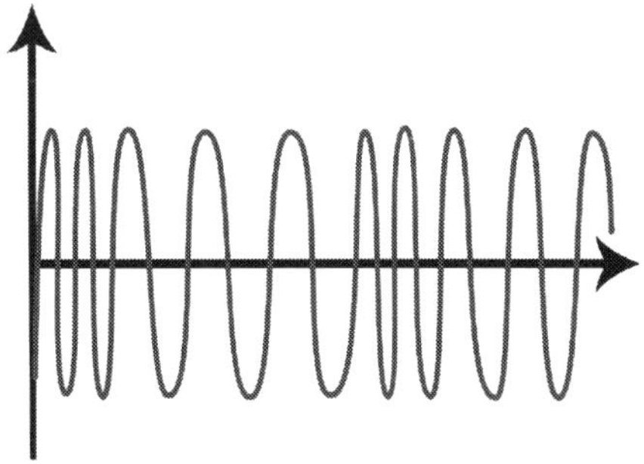

Practice Exam A
Question 049

4.08	5.18	7.02

Consider a 9-blade fan experiencing 12.5 mm/sec 0-p (or 0.5 ips 0-p) vibration. The fan runs at an unknown speed. If the problem is mass unbalance, what is the vibration frequency in cpm and order in this case? RS stands for running speed.

	Frequency	
	cpm	Order
A.	12.5	9X RS
B.	1795	1X RS
C.	112.5	Unknown
D.	Unknown	1X RS

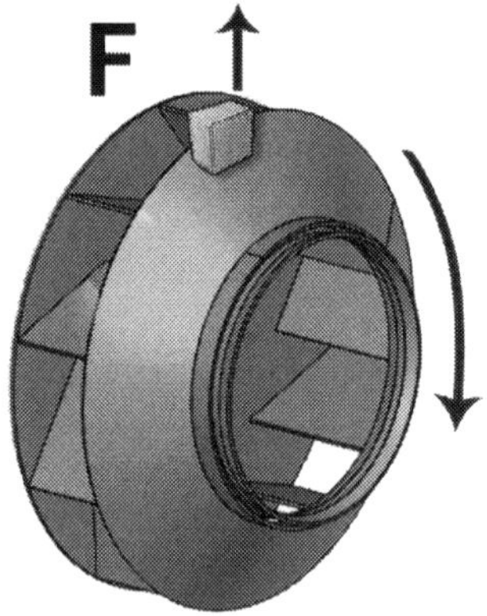

Practice Exam A

Question 050

Consider a typical centrifugal pump. What is a key symptom of shaft misalignment? Select the best answer from the given choices.

A. The main frequency components are usually at 1X and 2X running speed.

B. The main frequency components are usually at 1X, ½ X, 1/3 X and ¼ X running speed.

C. The main frequency component is at running speed multiplied by the number of vanes.

D. Vibration spectrum shows noticeable non-integer orders of the shaft's running speed.

Note

Practice Exam A

Question 051

A small fan (1498 rpm) is mounted on rolling element bearings. The vibration collected in velocity from the bearing housings shows several peaks separated by 25 Hz. What is the most probable the fault? Refer to the next illustrations.

A. The information suggests a classic sign of looseness.

B. The information suggests a an early stage of an unbalance condition.

C. The information suggests a strong electrical grounding fault.

D. The information suggests an early stage of high resolution.

Note

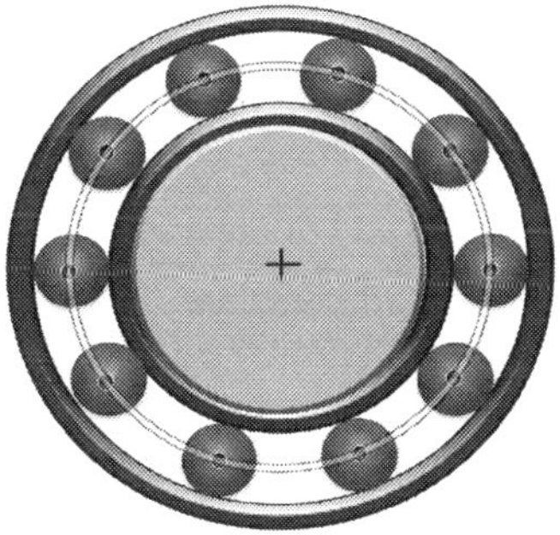

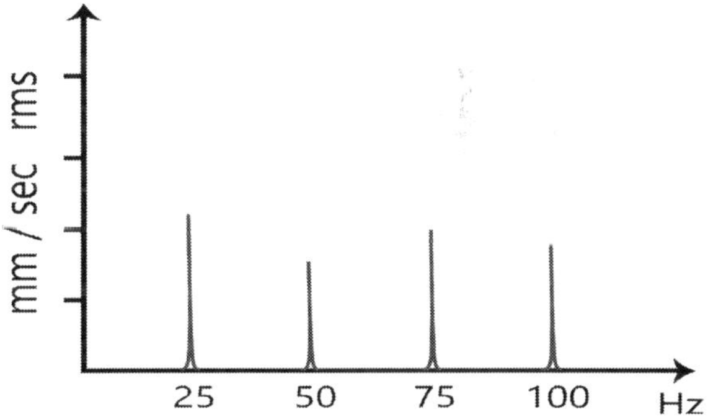

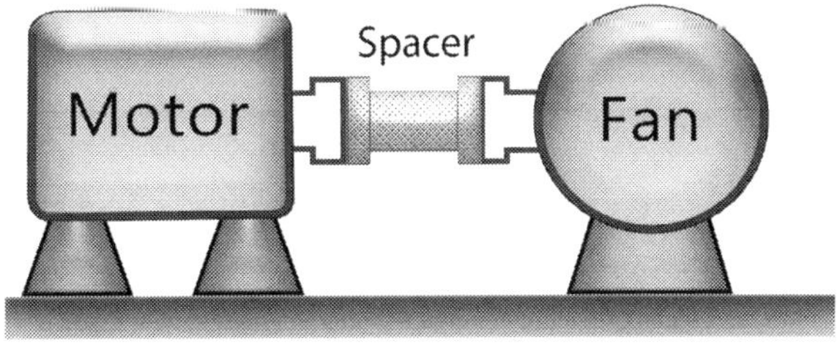

Note

Practice Exam A
Question 052

1.04	2.03	8.1	

Study the given table that is used to evaluate the severity of machinery vibration. Which of the following vibration cases is the most suitable case to be evaluated using Table 1?

A. Vibration collected from a very small motor (less than 4 Hp or 3 kW)

B. Vibration collected from a large motor monitored by online eddy current probes.

C. Vibration collected from a shaft of a large motor with journal bearings.

D. Vibration collected from a large motor monitored by online displacement probes.

Note

Table 1	Overall Vibration Severity Guide for operating Speed Faults (inch per second)
MACHINE SEVERITY	RMS
Acceptable	Less than 0.08
Normal	0.08 to 0.12
Surveillance	0.12 to 0.28
Unacceptable	More than 0.28

Note

Practice Exam A
Question 053

1.03	1.04	8.1	

What is the overall amplitude of the next vibration waveform taken from a centrifugal pump? What is the condition of the pump? Alarm = 4 mm/sec 0-p. Based on this limited information, select the best answer based on the provided information.

	Amplitude (mm/sec pk)	Condition
A.	1.2	Normal
B.	2.1	Alarm
C.	15	Dangerous
D.	30	Unknown

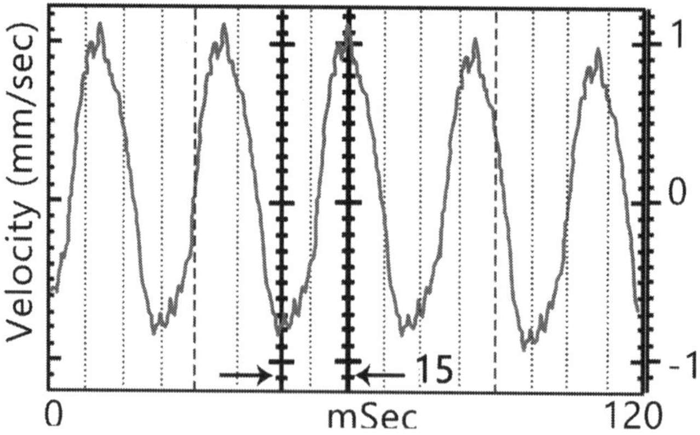

1.03	1.05	8.1		Practice Exam A Question 054

The following trends are collected from sister pumps by the same online system. Which case is the worst in terms of vibration severity? All points show major 1X amplitudes in their spectra.

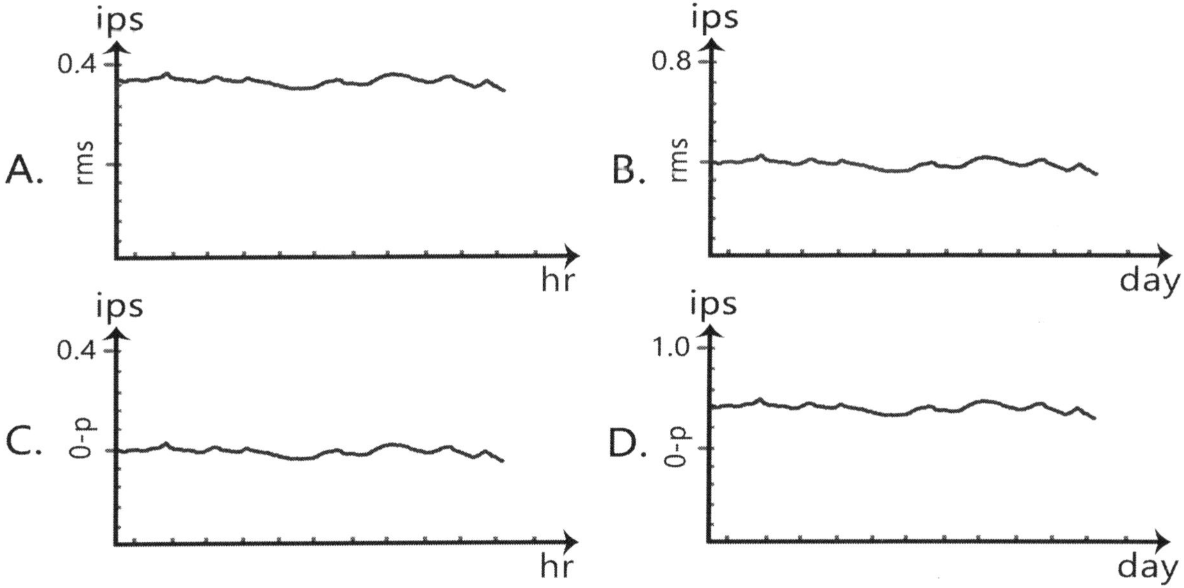

Practice Exam A

Question 055

8.1

What is the condition evaluation of a water pump (1790 rpm) whose vibration spectrum is below. High Alarm = 0.2 ips 0-p (5 mm/s 0-p), High High Alarm = 0.4 ips 0-p (10 mm/s 0-p).

A. Normal condition

B. Surveillance condition

C. Dangerous condition

D. Evaluation cannot be done without knowing the frequency

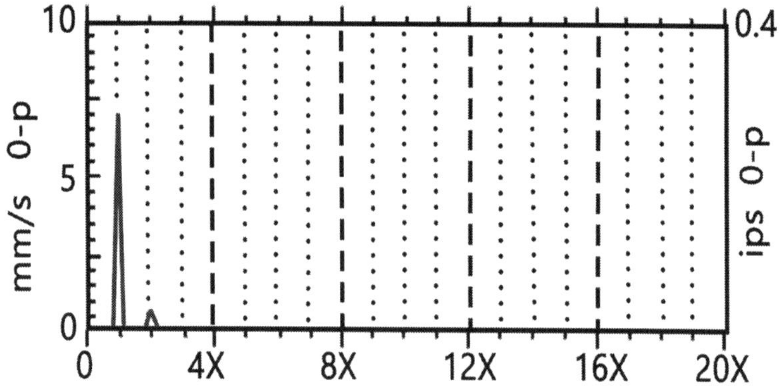

7.04

Practice Exam A
Question

Which part exists in typical centrifugal compressors but does not exist in typical screw compressors?

A. Seal

B. Impeller

C. Bearing

D. Shaft

Note

Practice Exam A

Question 057

8.1

Study the next illustration to evaluate the condition of the pump. What is your recommendation? All readings are in inch/sec peak. Use Table 3.

A. Keep running

B. Acceptable

C. Surveillance

D. Unacceptable

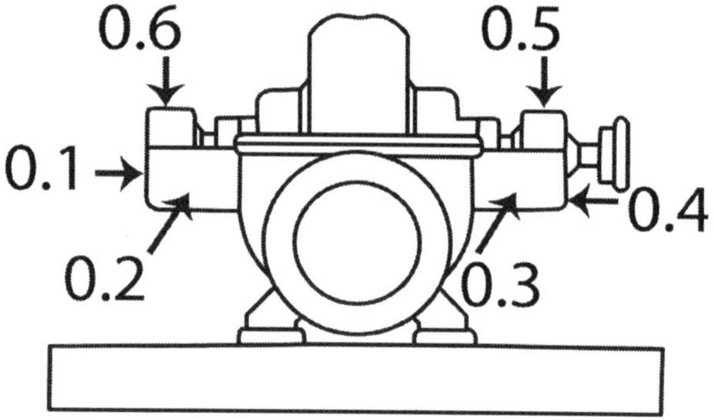

Note

Table 3 SEVERITY	Overall Vibration (inch/second peak)
Acceptable	Less than 0.30
Surveillance	0.30 to 0.7
Unacceptable	More than 0.7

Practice Exam A
Question 058

8.1

Study the next motor name plate. What is the vibration frequency of 2X? What is the frequency of electric power supply?

	2X	Electric Power Frequency
A.	25 Hz	450 V
B.	100 Hz	24.17 rpm
C.	1450 cpm	5.21 A
D.	2900 cpm	3000 cpm

Note

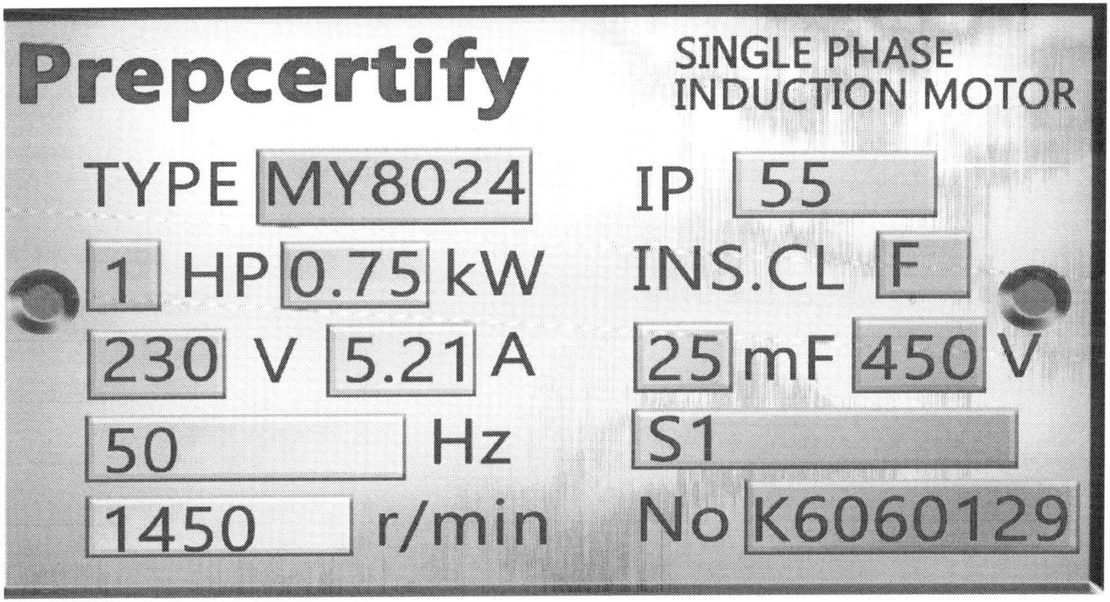
Prepcertify
SINGLE PHASE
INDUCTION MOTOR
TYPE MY8024
IP 55
1 HP 0.75 kW
INS.CL F
230 V 5.21 A
25 mF 450 V
50 Hz
S1
1450 r/min
No K6060129

Practice Exam A

Question 059

4.08 | 5.18

Consider a pump running at 1795 rpm. Which of the following spectra is the best indication of mechanical looseness? The amplitude is given in inch per second.

A.

B.
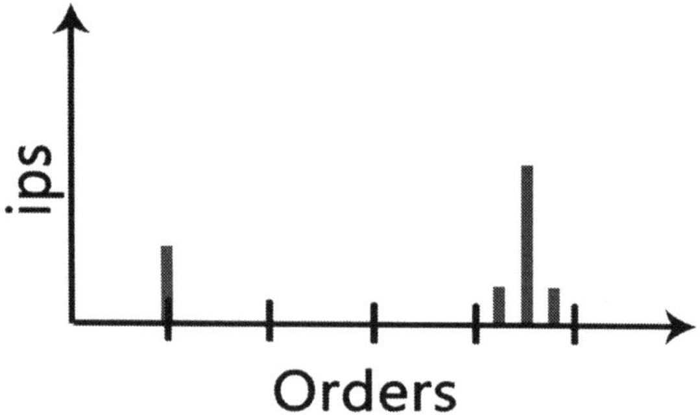

C.
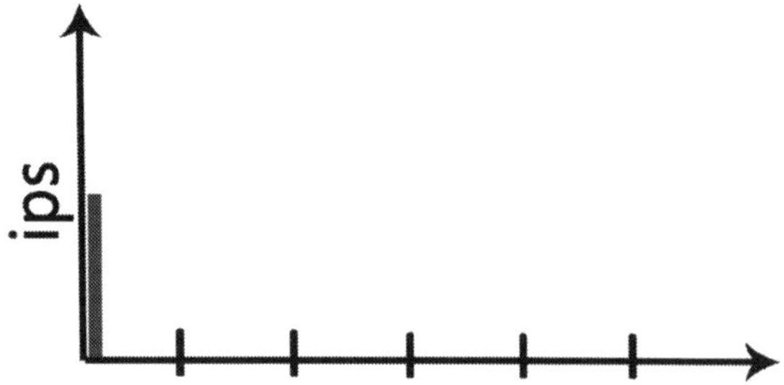

D.
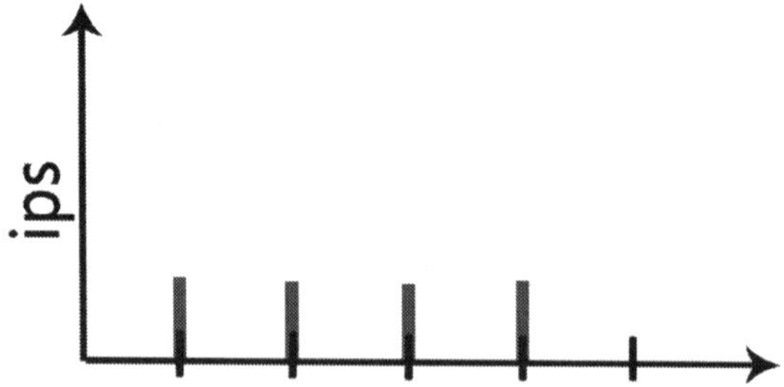

1.03 8.1

Practice Exam A
Question

Study the next unfiltered vibration data of a large pump (about 1000 hp, 1270 rpm). What is the pump's condition provided that the set points are Alarm = 2.4 mils pp and Danger = 3.6 mils pp.

A. Acceptable

B. Warning

C. Danger

D. Condition is unknown

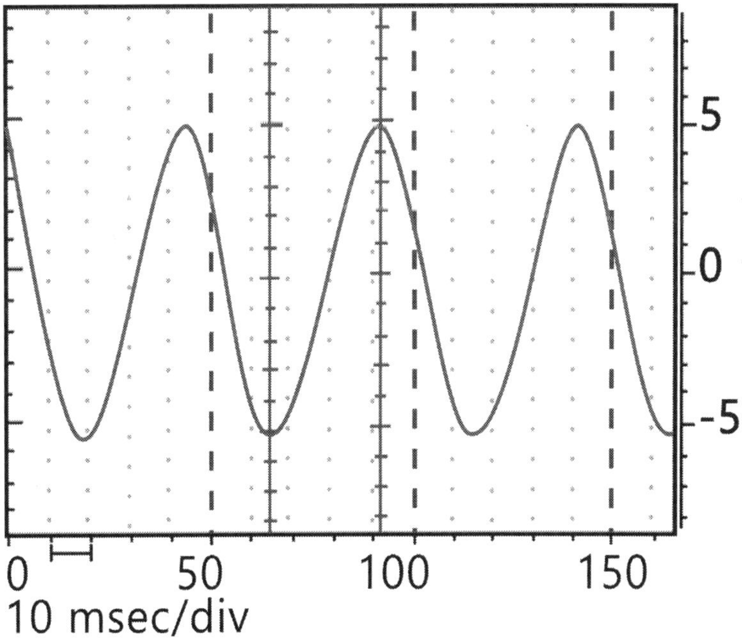

Notes

120
Questions
PART
8
2022

EXAM B

TWO PRACTICE EXAMS

Ali M. Al-Shurafa

Practice Exam B

Question 001

3.05

What type of vibration plot is in the below illustration? The data was collected from a small gear pump.

A.	Gap voltage trend in zero to peak	C.	High resolution spectrum
B.	Waveform without a phase reference	D.	Casing overall vibration trend

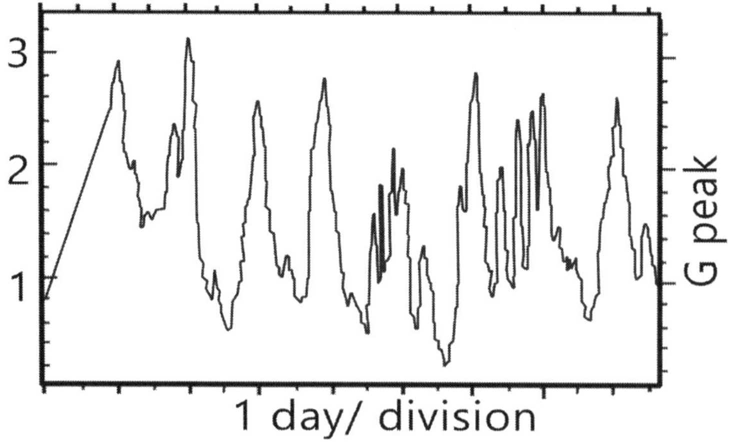

1.01 3.05

Question

Which plot type can be considered as the origin of the other vibration plots?

A. Gap voltage trend

B. Waveform

C. High resolution frequency spectrum

D. Ski slope

Note

Practice Exam B

Question 003

1.02 | 1.06 | 3.5

Which of the following plots is a vibration frequency spectrum?

A

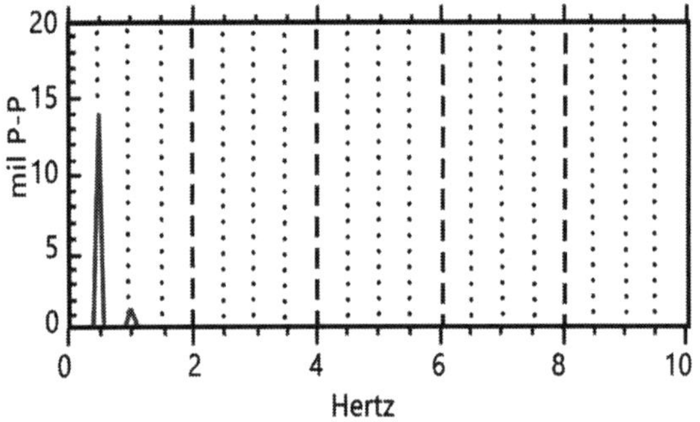

B

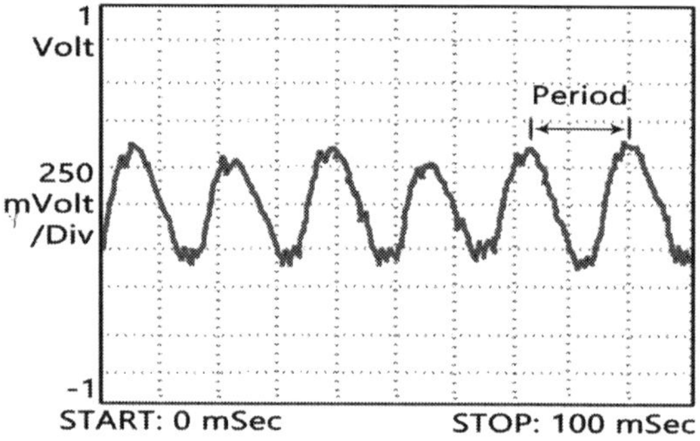

C

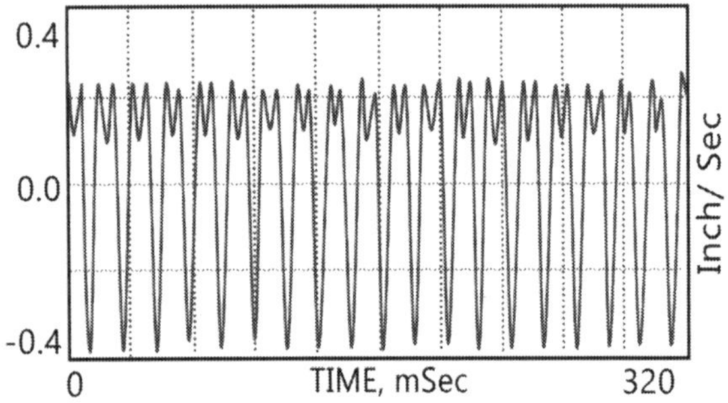

D

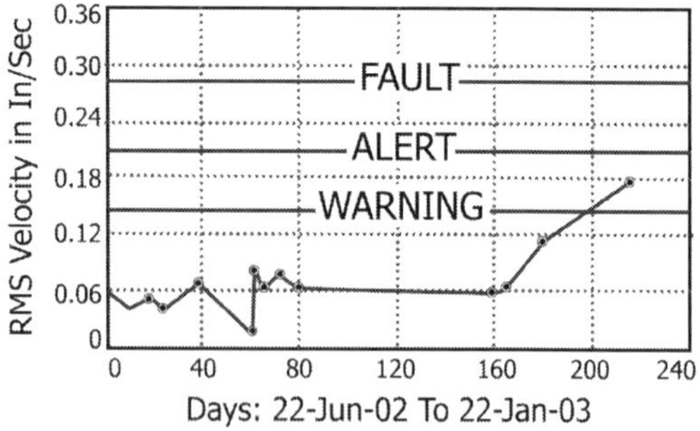

1.02

Practice Exam B
Question

A vibration signal repeats itself every 4.5 msec. What is the frequency in Hz? Expect some rounding.

A. 0.22

B. 222

C. 4,500

D. 13,333

Note

Practice Exam B

Question 005

1.02

Study the next plot. What is the period in seconds? Expect some rounding.

A. 4.5

B. 20

C. 2.8

D. 0.022

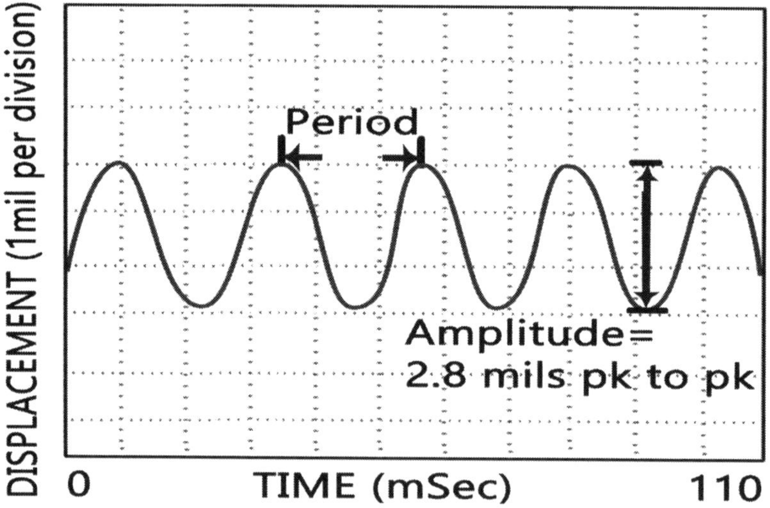

1.01 1.02 3.05

Practice Exam B
Question

What is the frequency and the period of the following vibration waveform signal?

	Frequency	Period
A.	0.1 Hz	0.1 msec
B.	0.1 sec	100 msec
C.	10 Hz	0.1 sec
D.	None of the above.	

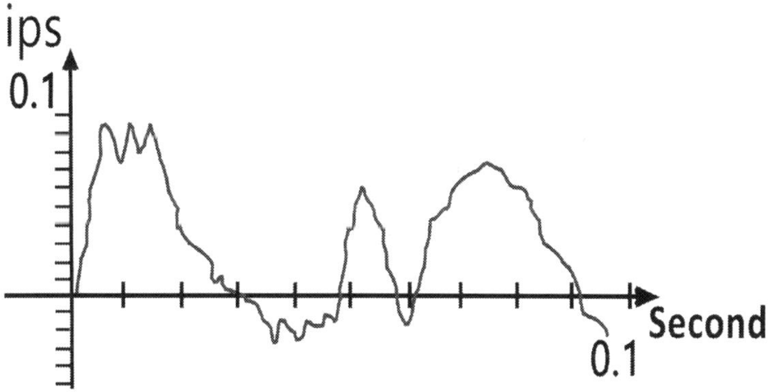

Note

Practice Exam B
Question 007

1.02

Study the provided vibration plot. What is the frequency of the dominant signal? Some answers are approximated. Select the best answer from the given choices.

A. 0.0169 sec

B. 60 cpm

C. 60 Hz

D. 0.017 Hz

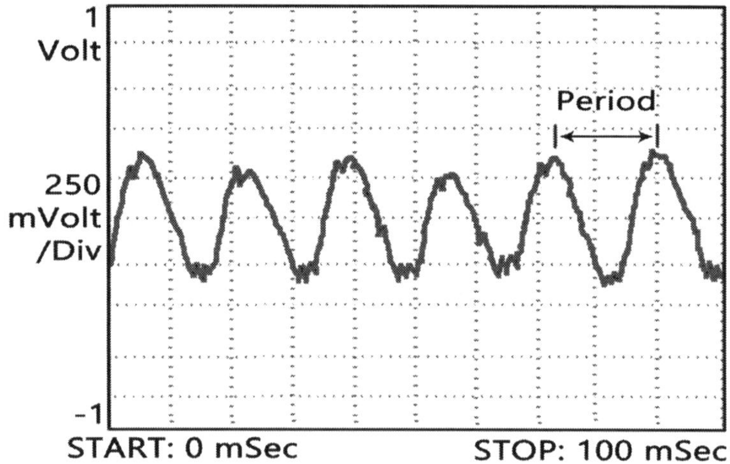

1.04

Practice Exam B
Question

What are the typical units used for the following parameters in the field of machinery vibration analysis? The amplitude SI units are provided in parenthesis.

	Displacement Amplitude	Velocity Amplitude
A.	mil (μm)	cpm
B.	mile (km)	rpm
C.	mil (μm)	in/sec (mm/sec)
D.	Hz	in/sec (mm/sec)

Note

Practice Exam B

Question 009

1.03

For the next plot, what is the amplitude in mils (p-p and rms)? Expect some rounding.

	P-P	RMS
A.	10	3.5
B.	10	7.07
C.	5	360
D.	5	3.5

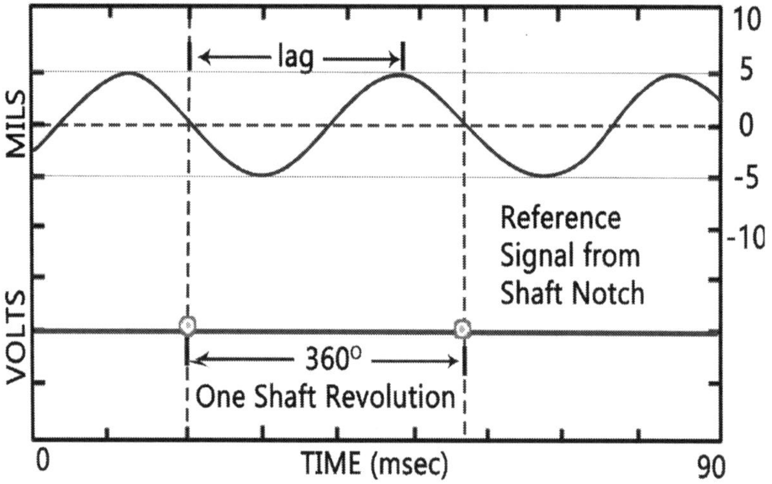

1.05

Practice Exam B
Question

For the next vibration spectrum, what is the 1X amplitude expressed in IPS peak? Assume the machine is running at 3599 rpm but under 105% load. Expect some number rounding.

A. 0.085

B. 0.18

C. 0.255

D. 0.128

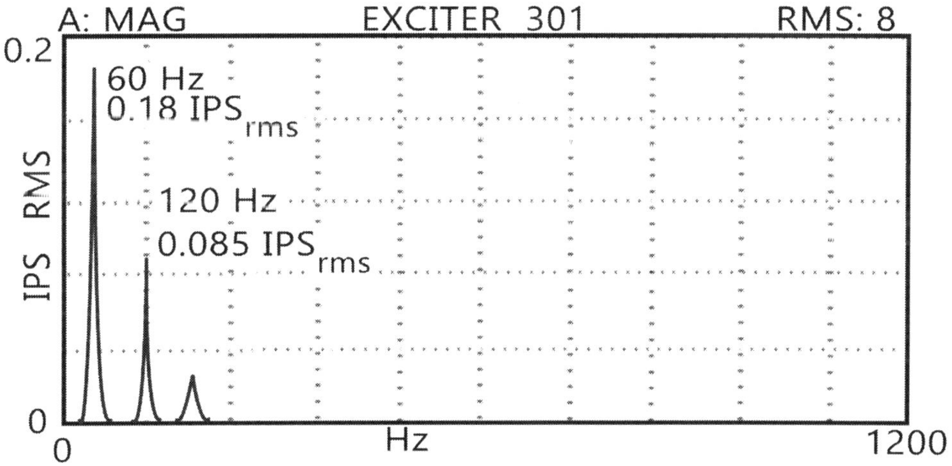

Practice Exam B

Question 011

1.04	1.05		

Convert the amplitude of a simple vibration signal from displacement mils pp to velocity ips rms. The overall vibration = 5 mils p-p which is mostly at 1X running speed.

A. 0.005 ips rms

B. 0.002 ips rms

C. 0.707 ips rms

D. The given information is not enough to perform the conversion.

Note

1.09	5.18	7.08

Practice Exam B
Question

What happens as the running speed of a machine and the machine's natural frequency get closer? Refer to the sketch.

A. Misalignment gets worse.

B. Mass unbalance increases.

C. Probability of a higher vibration increases.

D. Spectral resolution decreases.

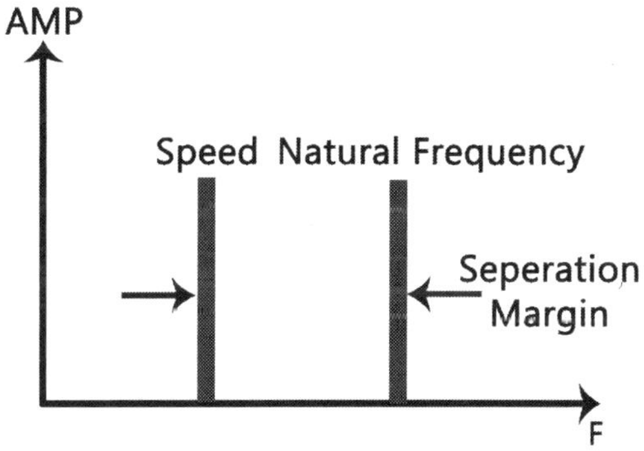

Note

Practice Exam B

Question 013

5.18

Study the next illustration. What is the phase lag between the triggering signal and the filtered vibration signal? Provide your answer in parts of revolution and degrees? Hint: Usually this value is referred to as 1X phase lag.

	Revolution	Degrees
A.	1/4 turns	270°
B.	3 turns	90°
C.	3/4 turns	270°
D.	3/4 turns	360°

Note

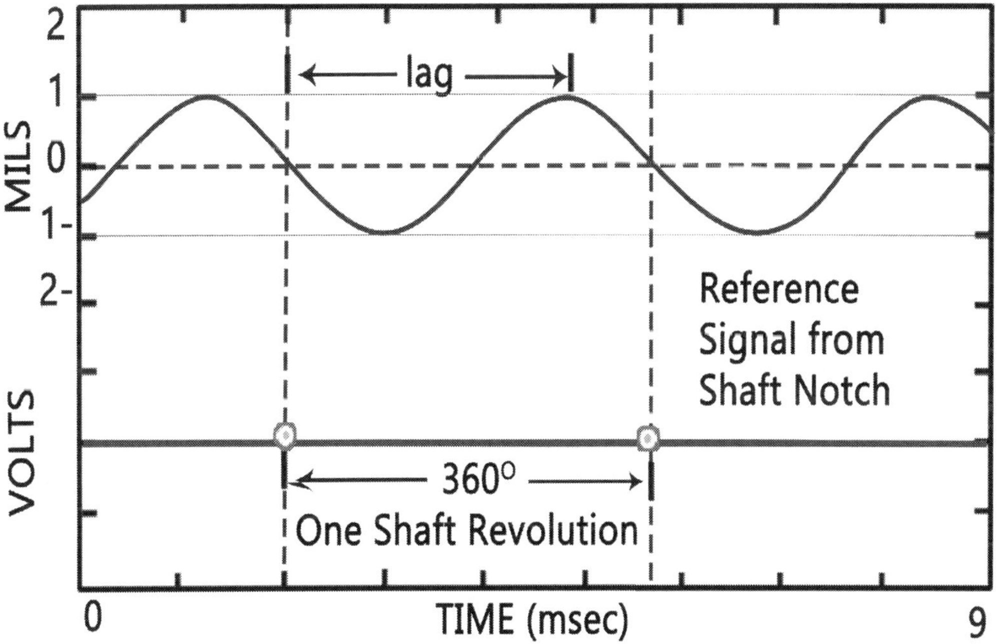
2
1
0
1-
2-
MILS
VOLTS
lag
Reference
Signal from
Shaft Notch
360°
One Shaft Revolution
0
TIME (msec)
9

Practice Exam B

Question 014

1.02 1.05

Study the next vibration frequency spectrum measured from a bearing housing. What is the frequency of the tallest vibration component? What is its order? Expect some rounding.

	Frequency	Order
A.	107 Hz	Insufficient information is given.
B.	107 Hz	1X
C.	6420 cpm	2X
D.	2.1 X	2.24 mm/sec rms

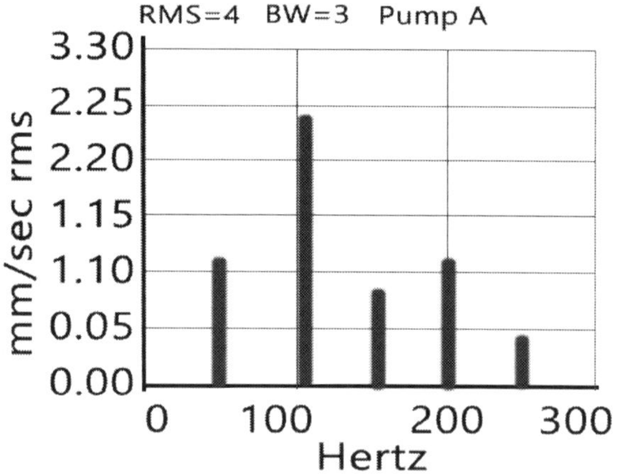

Note

Question

Study the next vibration spectrum. What is the frequency and amplitude for the main component?

A. 1.5 Hz and 0.100 ips rms

B. 120 Hz and 0.095 ips rms

C. 1X and 0.100 ips rms

D. 30 Hz and 2.48 mm/sec rms

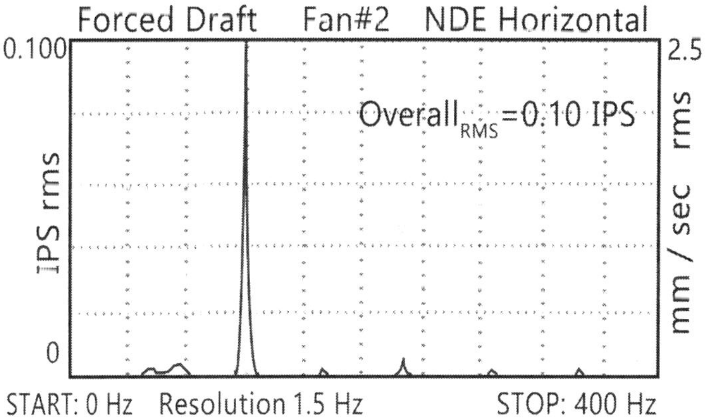

Note

Practice Exam B
Question 016

1.02 1.05

Study the next piping vibration plot. What is the frequency order indicated by P#3? The pump is running at 3000 rpm and data collection time is 0.8 seconds. Hint: convert the speed to Hz then locate 1X running speed component.

A. 2X

B. 2.4X

C. 4X

D. 6X

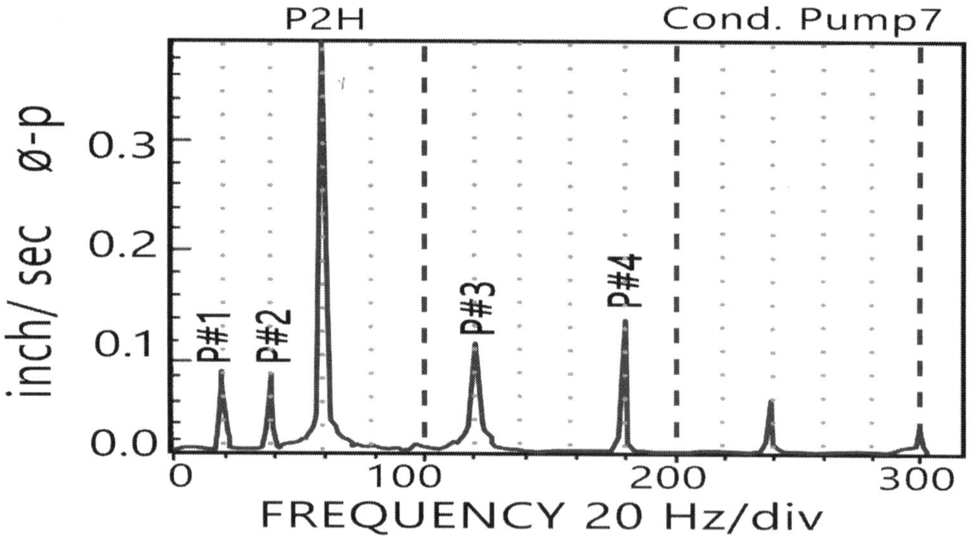

1.04 2.03

Practice Exam B
Question

What could you call a vibration probe that directly measures displacement? Velocity? Acceleration? Select the best answer from the given choices.

	Displacement	Velocity	Acceleration
A.	Relative probe	Velocity transducer	Accelerometer
B.	Tachometer	Speedometer	Gravitator
C.	Unidirectional	Bidirectional	Triaxial
D.	Positioner	Speeder	Decelerator

Note

Practice Exam B

Question 018

2.09 2.11

For vibration data collection from a small motor, how much time is typically needed for an analyzer to take a spectrum from a bearing housing? An example is below for a motor outboard point taken in velocity.

A. 0.01-0.10 seconds

B. 0.1-10 seconds

C. 10-50 seconds

D. 1-5 minutes

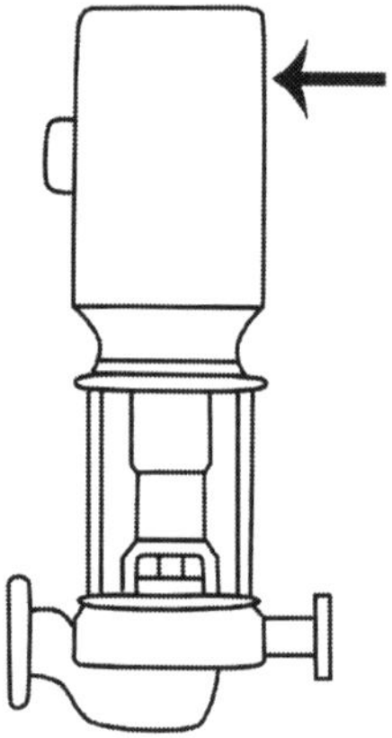

Note

2.09 8.1

Question

Which of the following is a safety concern when an analyst collects vibration data from rotating machines at a plant?

A. Cables of the data collector can be pulled by moving parts.

B. Analyst is exposed to hazards from the machine like high temperatures.

C. Attending equipment startup at the site exposes vibration analysts to risks if equipment fails.

D. All of the above

Note

Practice Exam B

Question 020

2.04 2.09 2.12

How to collect good quality and repeatable vibration measurements using a portable analyzer from small-medium size machines?

A. Use a magnet to hold the sensor on the bearing housings.

B. Measure on the same location every time.

C. Measure while machine is under the same/similar operating conditions.

D. All of the above.

Note

2.01 2.03 2.09

Practice Exam B

Question

An instrument technician reported that a vibration sensor has an acceptable gap of 50 mils (1.270 mm) after its re-installation on a compressor. Based on this limited information, what type of vibration-sensing element is this?

A. Relative probe that measures shaft vibrations.

B. Seismic probe that measures case displacement vibrations.

C. Velocity transducer that measures shaft vibrations.

D. Accelerometer that measures the gap between the shaft and its case.

Note

Practice Exam B
Question 022

2.03	2.04		

Which illustration is the best description for "stud mount" for an accelerometer?

A.
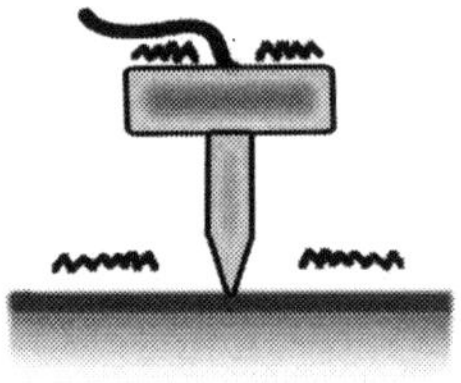

B.
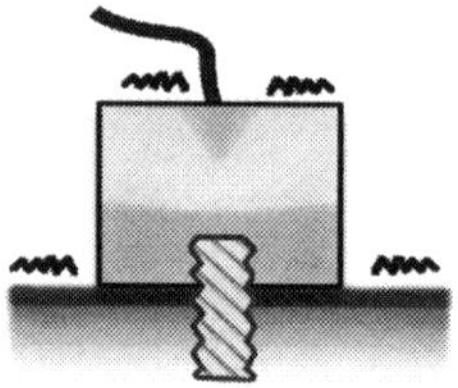

C.

D.
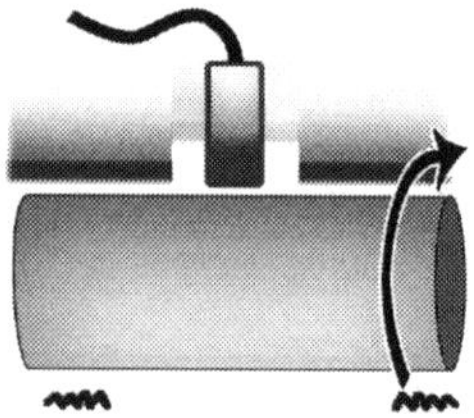

Note

2.09 7.02 8.01

Practice Exam B
Question

Refer to the next illustration for a single stage centrifugal pump. Which measurement point is referred to as "pump free end axial"? "Pump inboard vertical"?

	Pump Free End Axial	Pump Inboard Vertical
A.	1	5
B.	1	6
C.	4	2
D.	2	5

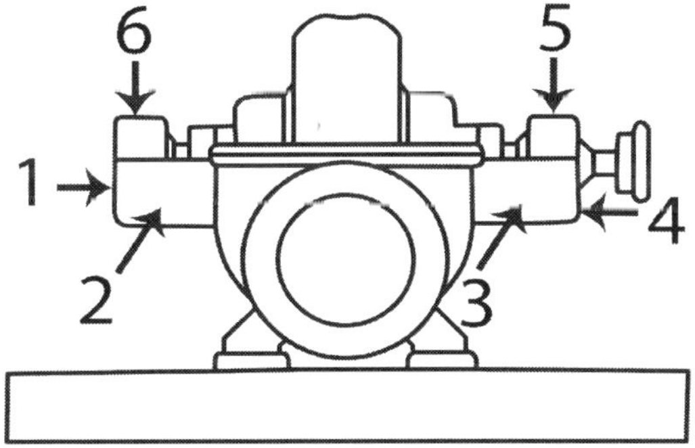

Note

Practice Exam B

Question 024

2.09	7.01	8.1	

Refer to the next illustration for a vertical motor driving a pump. This machine is covered in a vibration survey. Which point (indicated by the arrow) is referred to as "Motor Drive End Horizontal"?

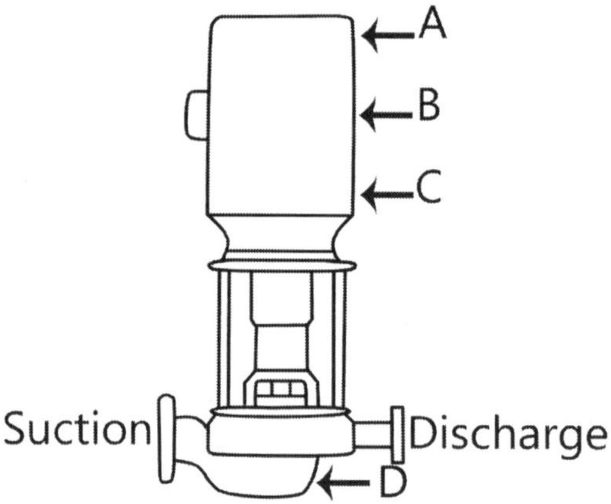

Note

2.03 2.04

Which of the following does not belong to the group of terms?

A. Magnet

B. Stud

C. Stinger

D. Accelerometer

Note

Practice Exam B
Question 026

2.03	2.09	8.1	

Consider a rotating machine with rolling element bearings and gears running at 1000-3600 rpm. Which of the following measurements are the best for evaluating the vibration's severity? Select the best choice from the given answers.

A. Shaft displacement and shaft velocity

B. Shaft velocity and shaft acceleration

C. Case displacement and case velocity

D. Case velocity and case acceleration

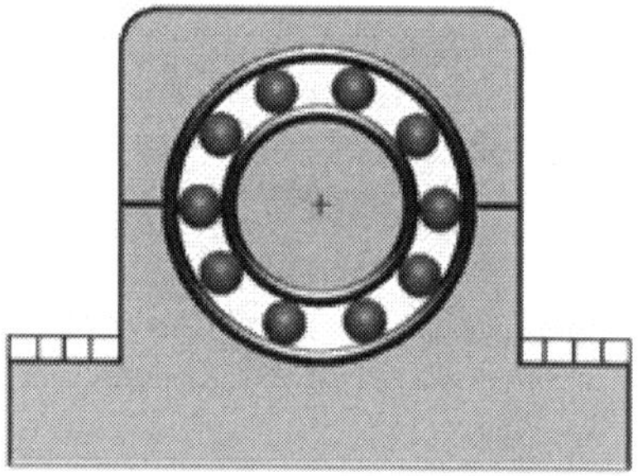

Note

2.01

Practice Exam B
Question

In the field of vibration data collection, what is true about single channel and multi-channel data collectors?

	Single channel data collector ________	Multi-channel data collector ________
A.	can collect a waveform.	can collect waveform and orbit.
B.	can collect waveform and orbit.	can collect a waveform.
C.	is usually used for small- medium machines.	is usually used for large machines.
D.	A and C	

Note

Practice Exam B

Question 028

2.11	3.05	8.1	

Assume you want to configure your analyzer to take vibration data from a machine. In general, what is the main factor that determines the correct value of F_{max} set in the analyzer? Select the best answer from the given choices.

A. The frequencies of the expected faults (of the machine to be surveyed).

B. The time given to the analyst to collect vibration data (for each machine).

C. The available memory in the analyzer.

D. The angular location (phase angle) of mass unbalance on the rotor.

Note

2.09 2.12 8.1

Practice Exam B

Question

If the overall vibration reading from a bearing housing is different every minute when the data is collected, what could be a reason for this instability?

A. Significant process (operational) upset

B. Faulty instrumentation device

C. Background vibration

D. All of the above

Note

Practice Exam B

Question 030

3.05

The purpose of FFT algorithm is to transform vibration signals from ____________. (Complete).

A. the probe to the control room.

B. a high to low amplitude.

C. the time domain to the frequency domain.

D. the frequency domain to the time domain.

Note

2.09 2.12

Practice Exam B
Question

What are two descriptions for the below spectrum which is taken from a large motor? Select the best from the below expressions.

A.	Ski slope	Low range frequency spectrum
B.	Wide band noise	Displacement spectrum
C.	Distinct components	Velocity spectrum
D.	Side bands	Displacement spectrum

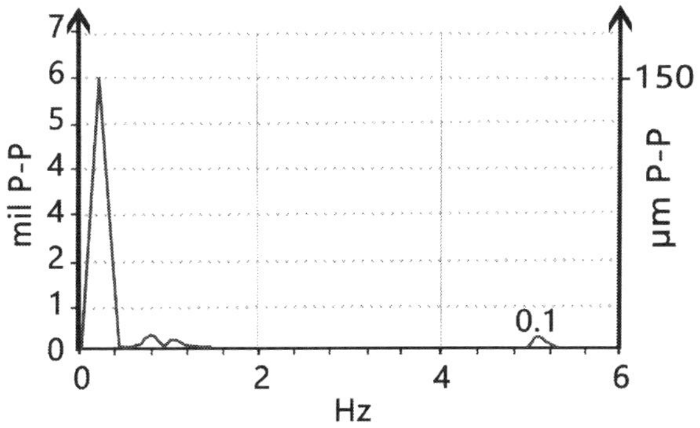

Practice Exam B
Question 032

2.09

What is the meaning of "calibration" in the field of machinery vibration?

A. Taking a second set of vibration data with a higher resolution.

B. Taking and averaging at least five (5) waveforms in order to produce a more reliable spectrum.

C. Verifying the accuracy of a measuring device based on a trusted reference through a defined process.

D. Remote data collection from critical machines using very accurate sensors.

Note

3.05

Practice Exam B
Question

Study the below vibration spectrum which was collected with 800 lines and displayed using a modern computer software. How many steps can the cursor move through the spectrum from 0 Hz to 200 Hz (from F_{min} to F_{max})?

A. 10 steps

B. 800 steps

C. 1024 steps

D. 16000 steps

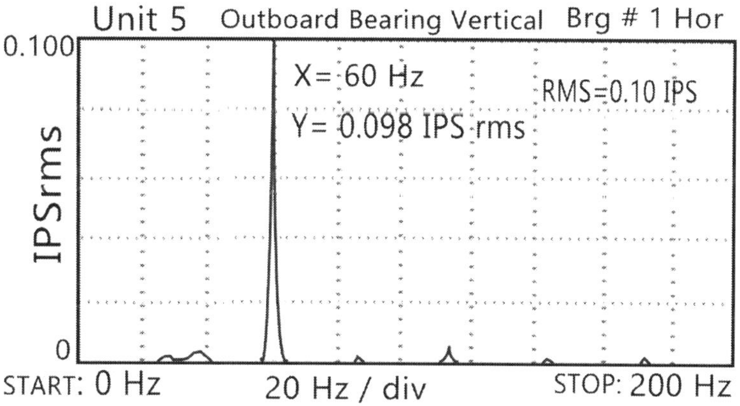

Practice Exam B

Question 034

1.05	1.06	3.05	

Study the next plot. What is F_{max} and the data acquisition time (DAT)?
Hint: Write the equation for DAT.

	F_{max} (Hz)	DAT
A.	60	0.098 msec
B.	60	16.7 msec
C.	200	No enough data is available
D.	200	0.005 seconds

Note

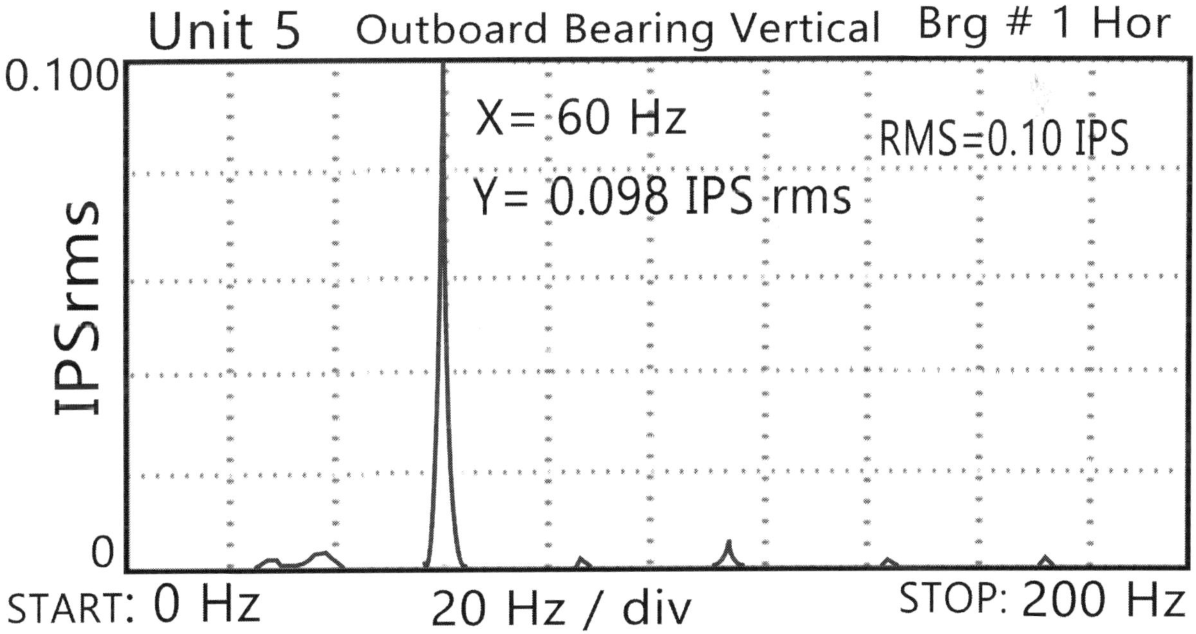
Unit 5
Outboard Bearing Vertical
Brg # 1 Hor
0.100
0
IPSrms
X= 60 Hz
Y= 0.098 IPS rms
RMS=0.10 IPS
START: 0 Hz
20 Hz / div
STOP: 200 Hz

Practice Exam B

Question 035

1.02	1.06	3.05

Study the next vibration waveform and spectra. Which spectrum is related the waveform?

A. 1

B. 2

C. 3

D. 4

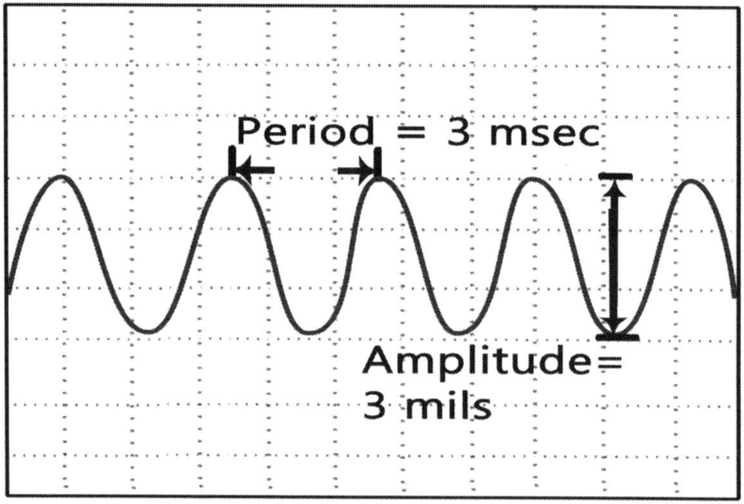

1.

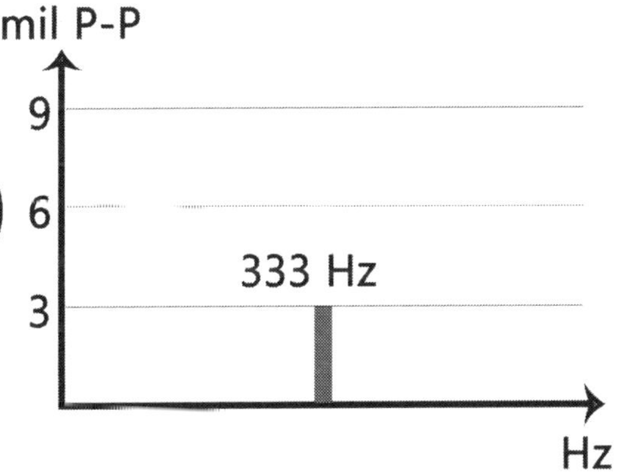
mil P-P
9
6
3
333 Hz
Hz

2.

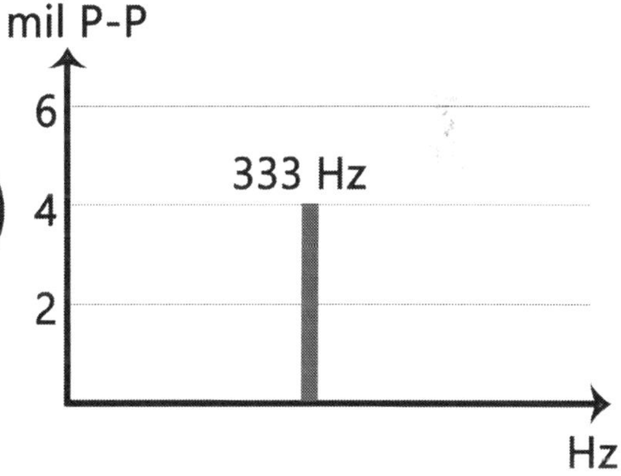
mil P-P
6
4
2
333 Hz
Hz

3.

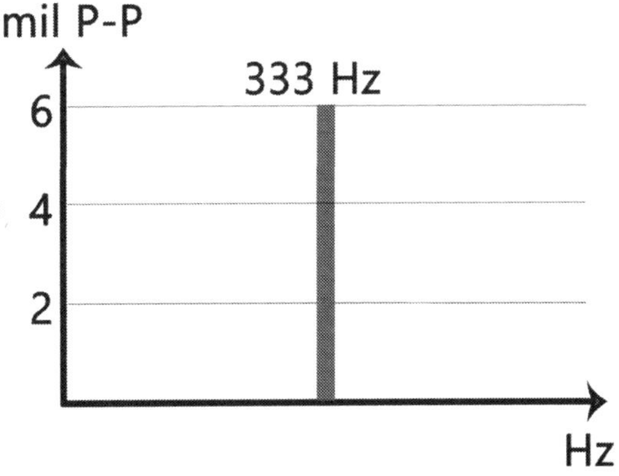
mil P-P
6
4
2
333 Hz
Hz

4.

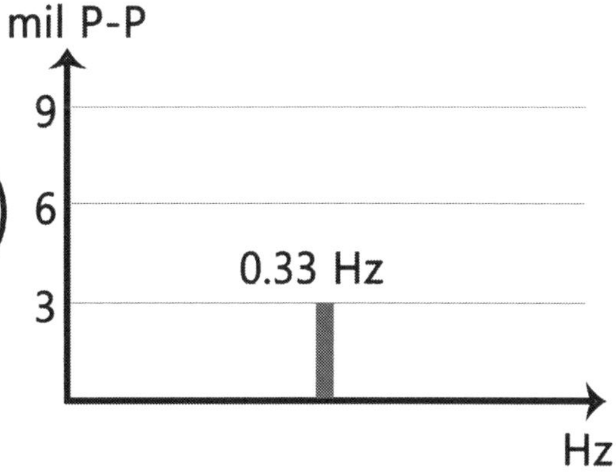
mil P-P
9
6
3
0.33 Hz
Hz

Practice Exam B

Question 036

3.05

True or False? In order to improve the resolution of a vibration spectrum, an analyst needs to increase both Frequency Span (F_{max}) and Lines of Resolution (LoR).
Hint: Write the equation of resolution.

A. True. Both must be increased in order to improve the resolution. This is the only option available for the analyst.

B. True. However, as an alternative, both F_{max} and LoR can be reduced in order to improve the resolution.

C. False. Increasing F_{max}, in fact, worsens the resolution.

D. False. Increasing LoR, in fact, worsens the resolution.

6.07 7.02 7.04

Practice Exam B
Question

What are the two most common types of bearings used for holding shafts in rotating equipment?

A. Journal Bearing and Sleeve Bearing

B. Fluid Film Bearing and Rolling Element Bearing

C. Spur Bearing and Helical Bearing

D. Rolling Element Bearing and Anti-friction Bearing

Note

Practice Exam B

Question 038

7.09

Refer to the next simplified illustration for a gearbox. What is the part indicated by 2?

A. GMF sensor

B. Pinion

C. Low speed hub

D. Thrust disk

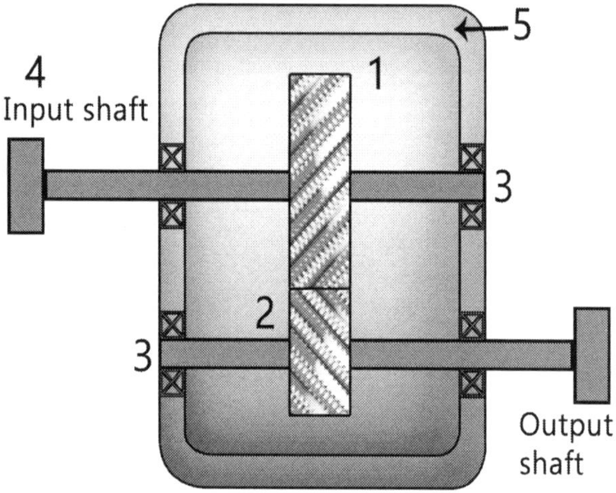

Note

7.09

Practice Exam B
Question

Study the next simplified illustration for a gearbox whose shafts are supported by rolling element bearings. Input speed is 1500 rpm. Which statement is true?

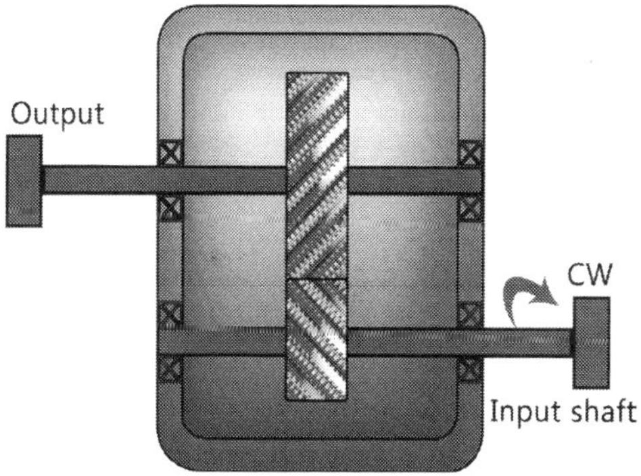

A. The speed of output shaft is higher than 1500 rpm.

B. The direction of rotation of the output shaft is counter clockwise (CCW), using a common viewing convention.

C. This gearbox has two (2) gear mesh frequencies.

D. To detect the high frequency faults of this gearbox, F_{max} must be <1,500 cpm.

Practice Exam B
Question 040

7.09

What is the gear mesh frequency for the following simple gearbox? Input shaft has 23 teeth and turns at 1800 rpm while its output shaft has 57 teeth. What is the speed of the output shaft? Hint: Add the numbers to the illustration. GMF = $N_1 \times T_1 = N_2 \times T_2$

	GMF	Output Speed
A.	690 cpm	4460 rpm
B.	690 Hz	726 rpm
C.	1710 Hz	74.3 Hz
D.	41400 cpm	Cannot be found from the given information

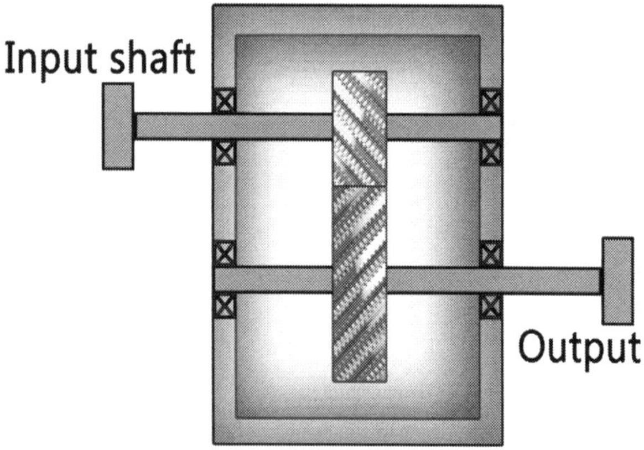

7.02 7.04

Practice Exam B
Question

What is true about a pump and a compressor?

	A pump ___________.	A compressor ___________.
A.	handles liquids.	handles gases.
B.	handles gases.	handles liquids.
C.	develops discharge pressures greater than or equal to 10 times suction pressure.	develops discharge pressures less than or equal to 10 times suction pressure
D.	develops discharge pressures less than or equal to 10 times suction pressure.	develops discharge pressures greater than or equal to 10 times suction pressure.

Practice Exam B

Question 042

7.04 7.06

There are many types of compressors. Which of the following are common types in the modern industry?

A. AC and DC

B. Centrifugal and positive displacement

C. Synchronous and induction

D. Forced and induced

Note

7.01

Practice Exam B

Question

Consider an AC three-phase, four-pole induction motor running on a 60 Hz line frequency with a 10 rpm slip. What is the 2-line frequency for this motor in Hz? What is the motor's frequency of 2X running speed? N = 2LF ÷ poles.

	2 Line Frequency (Hz)	2X Speed (rpm)
A.	60	3590
B.	120	3580
C.	1790	240
D.	7200	1790

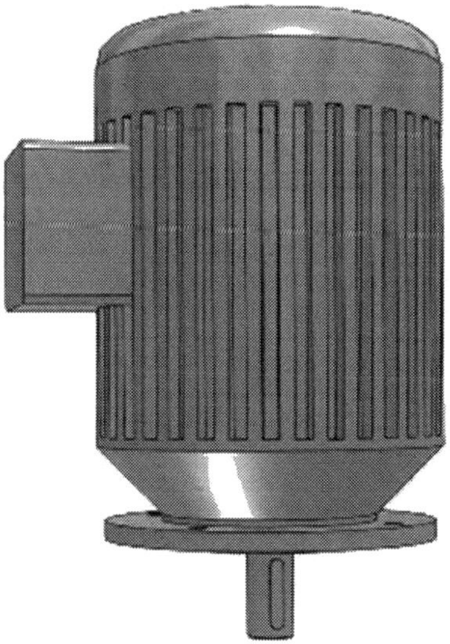

Practice Exam B
Question 044

7.04	7.06	8.1

Consider a classic small size reciprocating compressor. Which side has a higher temperature during normal operation: suction or discharge? refer to the next figure.

A. Suction

B. Discharge

C. They are equal.

D. It depends on the separation margin between the operating speed and critical speed.

Note

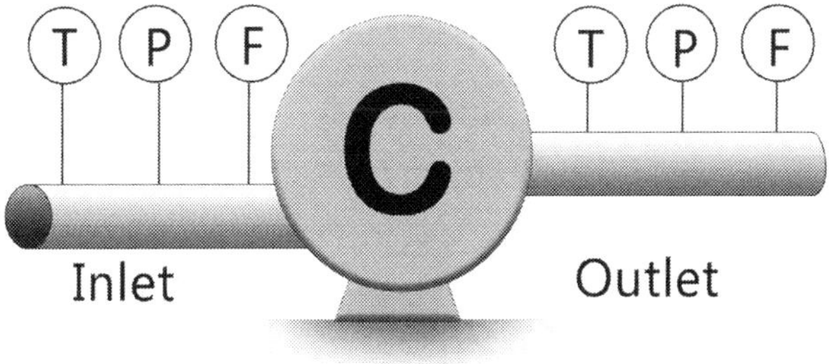
T
P
F
C
T
P
F
Inlet
Outlet

Practice Exam B

Question 045

For typical centrifugal pumps, cavitation can produce a high vibration and an audible noise. What is a common cause for cavitation? Select the best answer from the given choices.

A. Low suction pressure (below the design limit)

B. Low F_{max} during data collection (below the fault frequency)

C. Low flow rate

D. Poor vibration sensor installation

Note

7.06

Practice Exam B
Question

What is a key symptom of bearing damage noticed during vibration analysis? Consider a rolling element bearing installed in a rolling mill.

A. A gradual increase in the Data Collection Time.

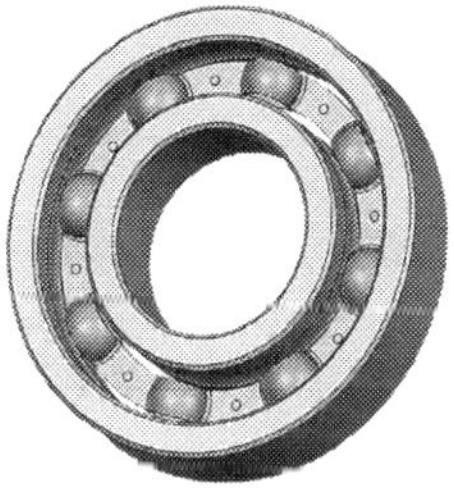

B. The main frequency components are at 1X, ½ X, 1/3 X and ¼ X running speed.

C. Vibration spectrum shows very high synchronous components.

D. Vibration spectrum shows noticeable non-synchronous components.

Note

Practice Exam B
Question 047

7.02 5.18

Consider a high resolution spectrum of a centrifugal pump (whose shaft is mounted on ball bearings). The vibration spectrum shows a low 1X amplitude and a high 5X amplitude. What could be the most probable problem? Based on this limited information, select the best answer from the given choices. Hint: Sketch the spectrum.

A. Bearing defect

B. Shaft misalignment

C. Hydraulic problem like cavitation or recirculation

D. Coupling unbalance

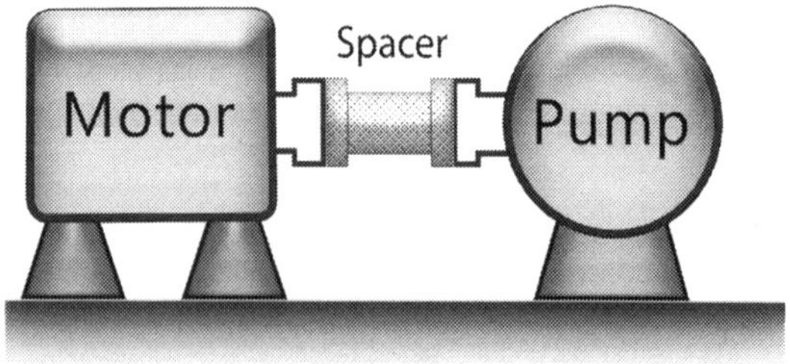

4.08 5.18

Practice Exam B
Question

Consider a pump (4 vanes) running at 1795 rpm. Which simplified spectrum from the following is the most indicative of mechanical looseness?

A.
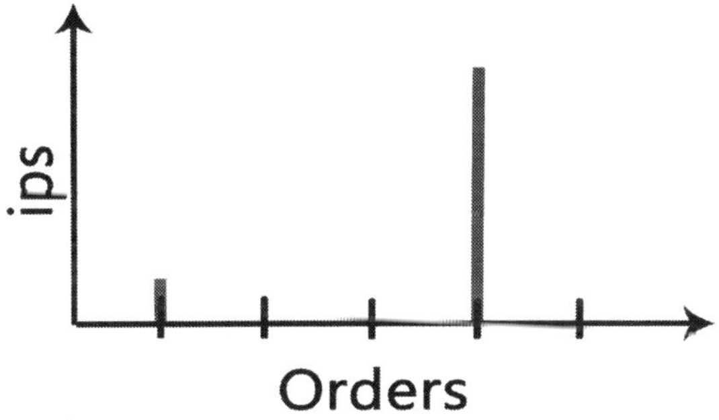

C.

B.
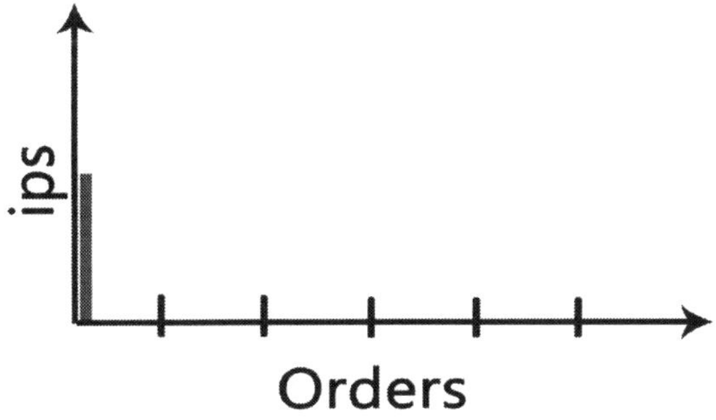

D.

Practice Exam B

Question 049

4.08 | 5.18

Consider a paper machine running at 600 rpm. Which of the following simplified spectra is the most possible signature of a rotor unbalance?

A.

B.

C.

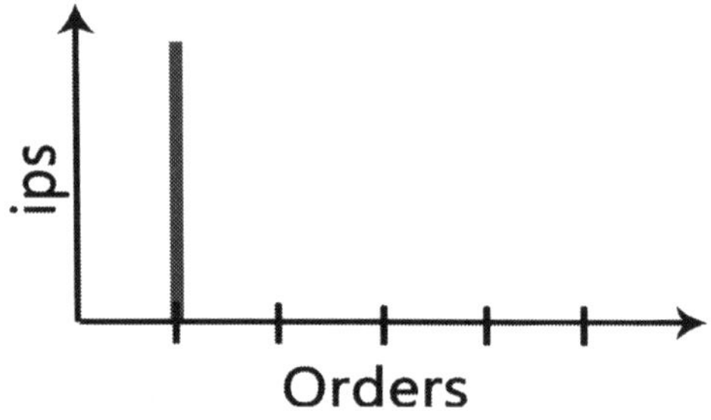

D.

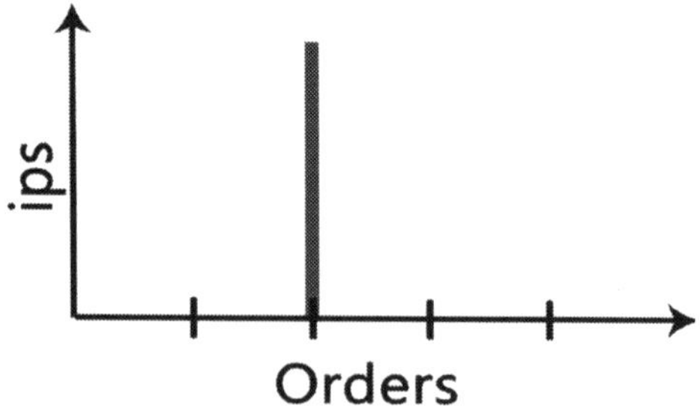

4.08 5.18 7.01

A vertical pump is driven by a four pole motor (1790 rpm). The unit has a natural frequency at 1812 cpm. The pump vibrates severely at the shaft running speed (1X) even after the motor was balanced in the shop. What could be the most probable cause of this vibration? Refer to the next illustrations.

A. Poor spectrum resolution

B. Structural resonance

C. Soft foot

D. High suction pressure

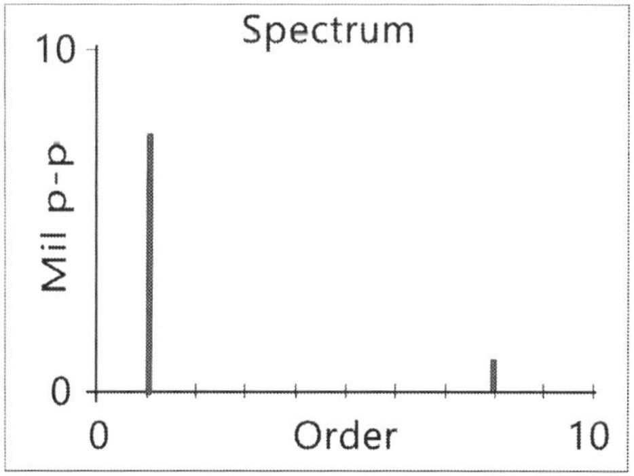

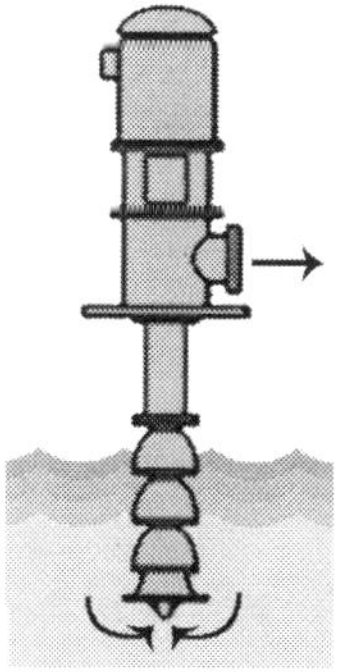

Practice Exam B
Question 051

4.08 | 5.18

Which waveform is the most probable indication for a rotor mass unbalance for a fan running at its full speed? Hint: pay attention to the time axis.

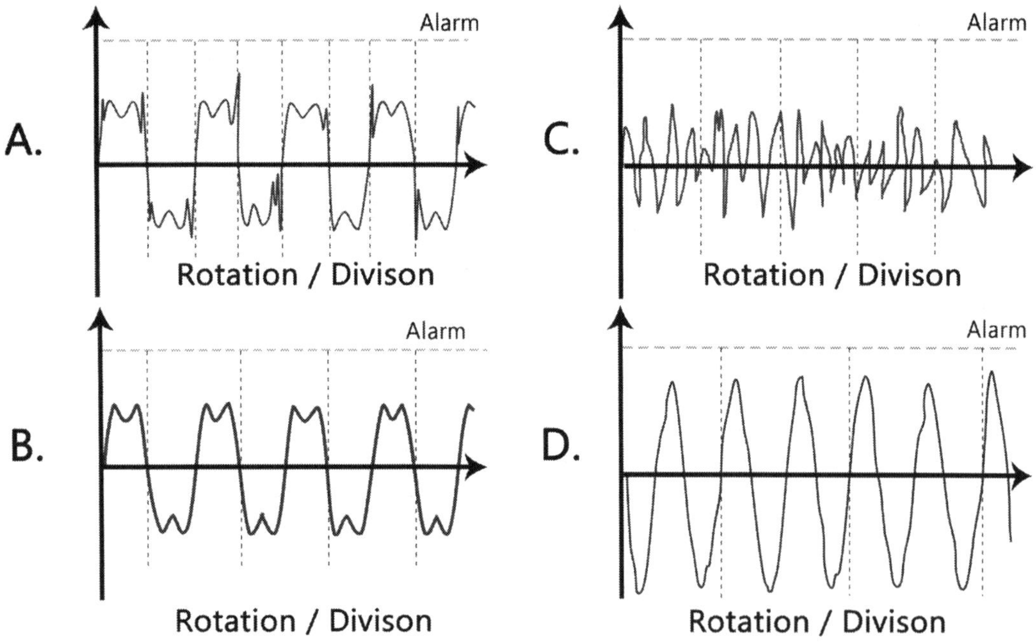

4.08 5.18 7.01

Practice Exam B
Question

What is a key vibration symptom that indicates a possible shaft misalignment problem? Consider a typical horizontal medium size motor driving a pump. Select the best answer from the given choices.

A. 1X, 2X and/or 3X shaft speed radial vibration, sometimes accompanied by axial vibration.

B. High frequency vibration (very clear when measured in acceleration).

C. High amplitude at 2LF noticed especially near to the coupling.

D. Combination of sidebands and floor noise in the spectrum.

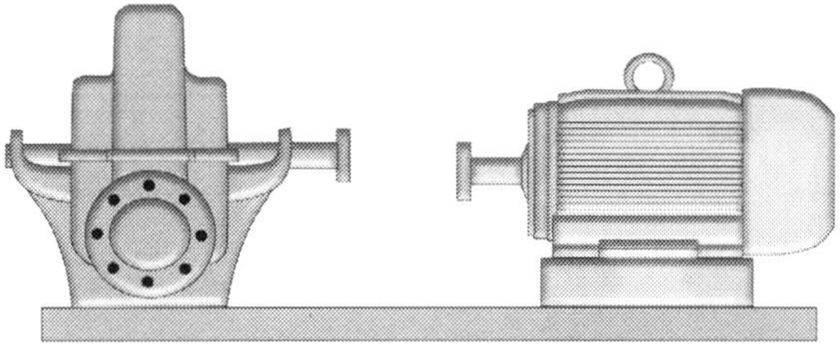

Practice Exam B

Question 053

1.03 | 8.1

Which waveform most likely demonstrates the worst vibration condition? Notice: 1 ips = 25.4 mm/s.

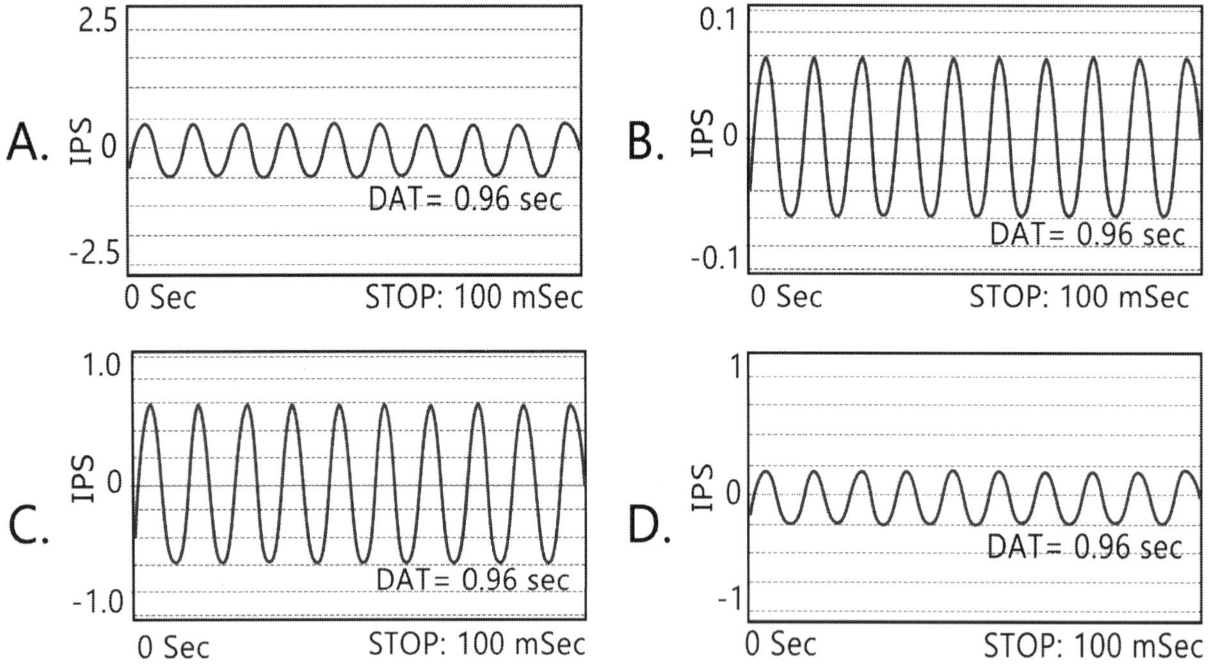

1.03 8.1

Practice Exam B
Question 054

The following seismic spectra are collected from sister machines in a power plant. Which of them is the worst in terms of vibration severity?

A.
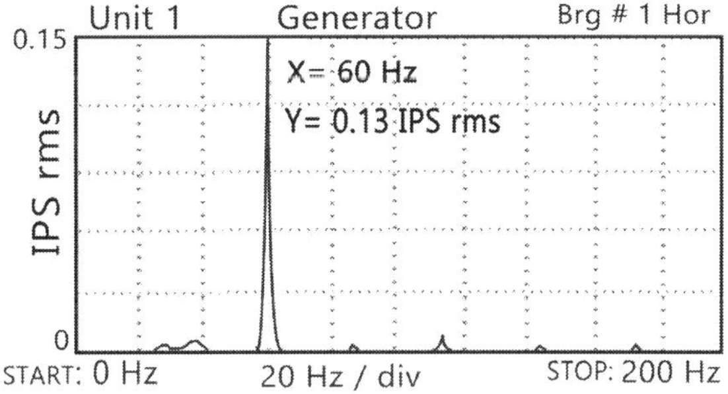

B.
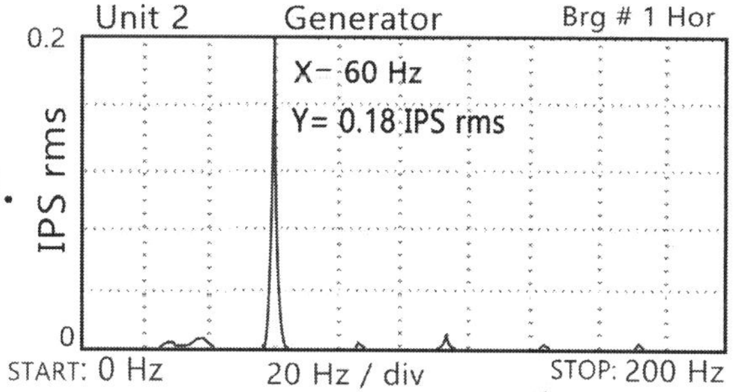

C.
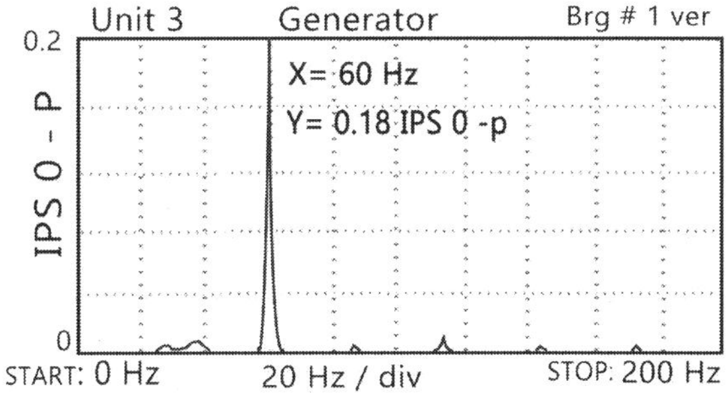

D.
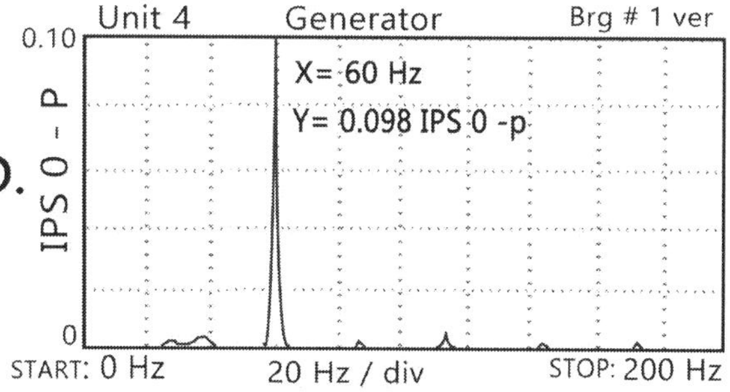

Practice Exam B

Question 055

2.3

What are the most common sensors used in the field of vibration analysis?

A.	Proximity probe system	Gear mesher	Alignment transmitter
B.	FFT sensor	Eddy current probe	Particle counter
C.	Relative probe	Velocity transducer	Accelerometer
D.	Pressure switch	Pressure transmitter	Pressure indicator

Note

2.09 2.11

Practice Exam B

Question

What is the most common period (cycle) for an in-house analyst to survey the machines for a routine data collection using a portable vibration analyzer? Consider classic equipment and industry.

A. 1-10 days

B. 2-8 weeks

C. 6-24 months

D. 2-4 years

Note

Practice Exam B
Question 057

2.09	7.01	8.1

Consider a classic horizontal medium size fan (center mounted) driven by a motor. How many vibration points are collected to achieve comprehensive but not excessive measurements? Assume you collect data in a route-based survey for both motor and fan. Refer to the next illustration.

A. 3

B. 4

C. 12

D. 18

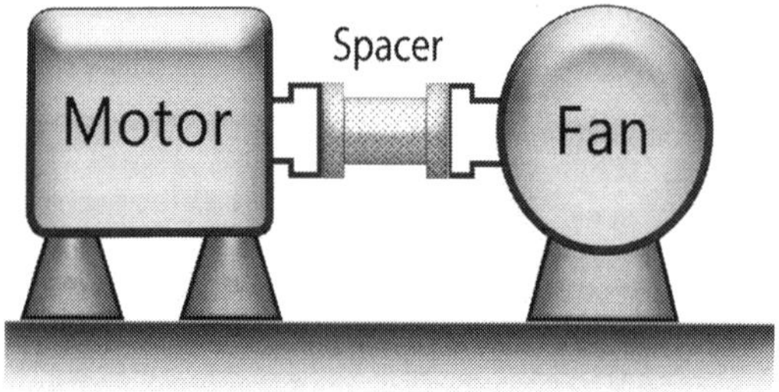

Note

7.9

Practice Exam B

Question

Study the next sketch for a simple gearbox. What is the output shaft's direction of rotation? What is the speed of the output shaft? Input teeth = 83, Output teeth = 18

Hint: $N_{in} \times T_{in} = N_{out} \times T_{out}$

	Direction	Output Speed
A.	CW	137 Hz
B.	CW	44297 rpm
C.	CCW	8203 rpm
D.	CCW	8203 Hz

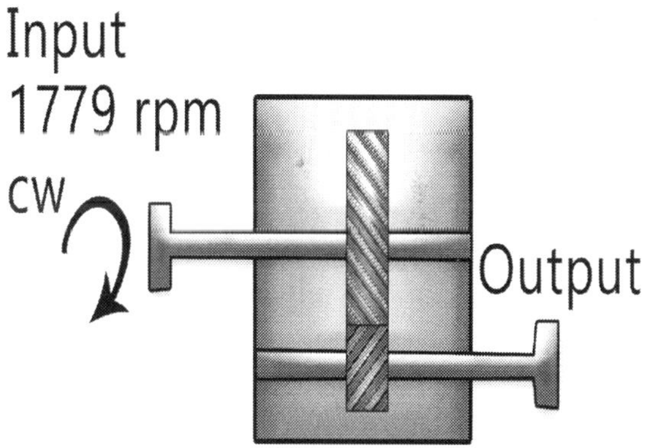

Note

Practice Exam B

Question 059 7.01 7.04

What is a very common kind of bearings usually used for large machines in the following industries: Power Generation, Chemicals, Petrochemicals and Oil & Gas?

A. Journal bearing

B. Grease lubricated bearing

C. Bevel bearing

D. Rolling element bearing

Note

2.03

Practice Exam B
Question

A 4-pole induction motor (1790 rev/min) drives a huge fan (2000 HP) through a simple gearbox. The pinion at the motor side has 29 teeth while the gear at the fan side contains 112 teeth. The shafts (of the three machines) are mounted on journal bearings. The power line is 60 Hz.

What is the gear mesh frequency? Expect some rounding in numbers.

A. 51.9 kHz

B. 1774 Hz

C. 865 Hz

D. 871 cpm

Notes

PRACTICE EXAM A
ANSWER SHEET 1

PRACTICE EXAM A EMPTY ANSWER SHEET 1

1	A	B	C	D
2	A	B	C	D
3	A	B	C	D
4	A	B	C	D
5	A	B	C	D
6	A	B	C	D
7	A	B	C	D
8	A	B	C	D
9	A	B	C	D
10	A	B	C	D
11	A	B	C	D
12	A	B	C	D
13	A	B	C	D
14	A	B	C	D
15	A	B	C	D

16	A	B	C	D
17	A	B	C	D
18	A	B	C	D
19	A	B	C	D
20	A	B	C	D
21	A	B	C	D
22	A	B	C	D
23	A	B	C	D
24	A	B	C	D
25	A	B	C	D
26	A	B	C	D
27	A	B	C	D
28	A	B	C	D
29	A	B	C	D
30	A	B	C	D

PRACTICE EXAM A EMPTY ANSWER SHEET 1

31	A	B	C	D
32	A	B	C	D
33	A	B	C	D
34	A	B	C	D
35	A	B	C	D
36	A	B	C	D
37	A	B	C	D
38	A	B	C	D
39	A	B	C	D
40	A	B	C	D
41	A	B	C	D
42	A	B	C	D
43	A	B	C	D
44	A	B	C	D
45	A	B	C	D
46	A	B	C	D
47	A	B	C	D
48	A	B	C	D
49	A	B	C	D
50	A	B	C	D
51	A	B	C	D
52	A	B	C	D
53	A	B	C	D
54	A	B	C	D
55	A	B	C	D
56	A	B	C	D
57	A	B	C	D
58	A	B	C	D
59	A	B	C	D
60	A	B	C	D

Notes

PRACTICE EXAM A
ANSWER SHEET 2

PRACTICE EXAM A EMPTY ANSWER SHEET 2

1	A	B	C	D
2	A	B	C	D
3	A	B	C	D
4	A	B	C	D
5	A	B	C	D
6	A	B	C	D
7	A	B	C	D
8	A	B	C	D
9	A	B	C	D
10	A	B	C	D
11	A	B	C	D
12	A	B	C	D
13	A	B	C	D
14	A	B	C	D
15	A	B	C	D

16	A	B	C	D
17	A	B	C	D
18	A	B	C	D
19	A	B	C	D
20	A	B	C	D
21	A	B	C	D
22	A	B	C	D
23	A	B	C	D
24	A	B	C	D
25	A	B	C	D
26	A	B	C	D
27	A	B	C	D
28	A	B	C	D
29	A	B	C	D
30	A	B	C	D

PRACTICE EXAM A EMPTY ANSWER SHEET 2

31	A	B	C	D
32	A	B	C	D
33	A	B	C	D
34	A	B	C	D
35	A	B	C	D
36	A	B	C	D
37	A	B	C	D
38	A	B	C	D
39	A	B	C	D
40	A	B	C	D
41	A	B	C	D
42	A	B	C	D
43	A	B	C	D
44	A	B	C	D
45	A	B	C	D

46	A	B	C	D
47	A	B	C	D
48	A	B	C	D
49	A	B	C	D
50	A	B	C	D
51	A	B	C	D
52	A	B	C	D
53	A	B	C	D
54	A	B	C	D
55	A	B	C	D
56	A	B	C	D
57	A	B	C	D
58	A	B	C	D
59	A	B	C	D
60	A	B	C	D

Notes

PRACTICE EXAM B
ANSWER SHEET 1

PRACTICE EXAM B EMPTY ANSWER SHEET 1

1	A	B	C	D	16	A	B	C	D
2	A	B	C	D	17	A	B	C	D
3	A	B	C	D	18	A	B	C	D
4	A	B	C	D	19	A	B	C	D
5	A	B	C	D	20	A	B	C	D
6	A	B	C	D	21	A	B	C	D
7	A	B	C	D	22	A	B	C	D
8	A	B	C	D	23	A	B	C	D
9	A	B	C	D	24	A	B	C	D
10	A	B	C	D	25	A	B	C	D
11	A	B	C	D	26	A	B	C	D
12	A	B	C	D	27	A	B	C	D
13	A	B	C	D	28	A	B	C	D
14	A	B	C	D	29	A	B	C	D
15	A	B	C	D	30	A	B	C	D

PRACTICE EXAM B EMPTY ANSWER SHEET 1

31	A	B	C	D
32	A	B	C	D
33	A	B	C	D
34	A	B	C	D
35	A	B	C	D
36	A	B	C	D
37	A	B	C	D
38	A	B	C	D
39	A	B	C	D
40	A	B	C	D
41	A	B	C	D
42	A	B	C	D
43	A	B	C	D
44	A	B	C	D
45	A	B	C	D

46	A	B	C	D
47	A	B	C	D
48	A	B	C	D
49	A	B	C	D
50	A	B	C	D
51	A	B	C	D
52	A	B	C	D
53	A	B	C	D
54	A	B	C	D
55	A	B	C	D
56	A	B	C	D
57	A	B	C	D
58	A	B	C	D
59	A	B	C	D
60	A	B	C	D

Notes

PRACTICE EXAM B
ANSWER SHEET 2

PRACTICE EXAM B EMPTY ANSWER SHEET 2

1	A	B	C	D
2	A	B	C	D
3	A	B	C	D
4	A	B	C	D
5	A	B	C	D
6	A	B	C	D
7	A	B	C	D
8	A	B	C	D
9	A	B	C	D
10	A	B	C	D
11	A	B	C	D
12	A	B	C	D
13	A	B	C	D
14	A	B	C	D
15	A	B	C	D

16	A	B	C	D
17	A	B	C	D
18	A	B	C	D
19	A	B	C	D
20	A	B	C	D
21	A	B	C	D
22	A	B	C	D
23	A	B	C	D
24	A	B	C	D
25	A	B	C	D
26	A	B	C	D
27	A	B	C	D
28	A	B	C	D
29	A	B	C	D
30	A	B	C	D

PRACTICE EXAM B EMPTY ANSWER SHEET 2

31	A	B	C	D
32	A	B	C	D
33	A	B	C	D
34	A	B	C	D
35	A	B	C	D
36	A	B	C	D
37	A	B	C	D
38	A	B	C	D
39	A	B	C	D
40	A	B	C	D
41	A	B	C	D
42	A	B	C	D
43	A	B	C	D
44	A	B	C	D
45	A	B	C	D

46	A	B	C	D
47	A	B	C	D
48	A	B	C	D
49	A	B	C	D
50	A	B	C	D
51	A	B	C	D
52	A	B	C	D
53	A	B	C	D
54	A	B	C	D
55	A	B	C	D
56	A	B	C	D
57	A	B	C	D
58	A	B	C	D
59	A	B	C	D
60	A	B	C	D

Notes

ANSWER KEY

ANSWER KEY EXAM A

Q. #	Answer	Q. #	Answer	Q. #	Answer	Q. #	Answer
1	A	16	B	31	A	46	A
2	A	17	A	32	A	47	B
3	D	18	A	33	B	48	B
4	D	19	B	34	C	49	D
5	A	20	B	35	B	50	A
6	B	21	B	36	A	51	A
7	C	22	C	37	A	52	A
8	D	23	D	38	B	53	A
9	D	24	C	39	A	54	D
10	B	25	C	40	A	55	B
11	D	26	D	41	B	56	B
12	D	27	A	42	D	57	C
13	C	28	D	43	A	58	D
14	D	29	D	44	D	59	D
15	D	30	B	45	D	60	C

ANSWER KEY EXAM B

Q. #	Answer	Q. #	Answer	Q. #	Answer	Q. #	Answer
1	D	16	B	31	A	46	A
2	B	17	A	32	C	47	C
3	A	18	B	33	B	48	D
4	B	19	D	34	C	49	C
5	D	20	D	35	A	50	B
6	D	21	A	36	C	51	B
7	C	22	B	37	B	52	A
8	C	23	A	38	B	53	C
9	A	24	C	39	B	54	B
10	C	25	D	40	B	55	C
11	D	26	D	41	A	56	B
12	C	27	D	42	B	57	C
13	C	28	A	43	B	58	C
14	A	29	D	44	B	59	A
15	B	30	C	45	A	60	C

Notes

ORDER OTHER PARTS OF CAT I PREP I PACKAGE

Don't guess where your skill stands; certify it. PrepCertify believes that the best preparation for professional certifications is obtained through practicing well-designed real world problems.

Learn what really matters in real world industry while mastering the Body of Knowledge in the certification standards. Your Cat I Prep I series does that for you. Through PrepCertify, you will achieve your certification in a much shorter time and with a greater value of your time and effort.

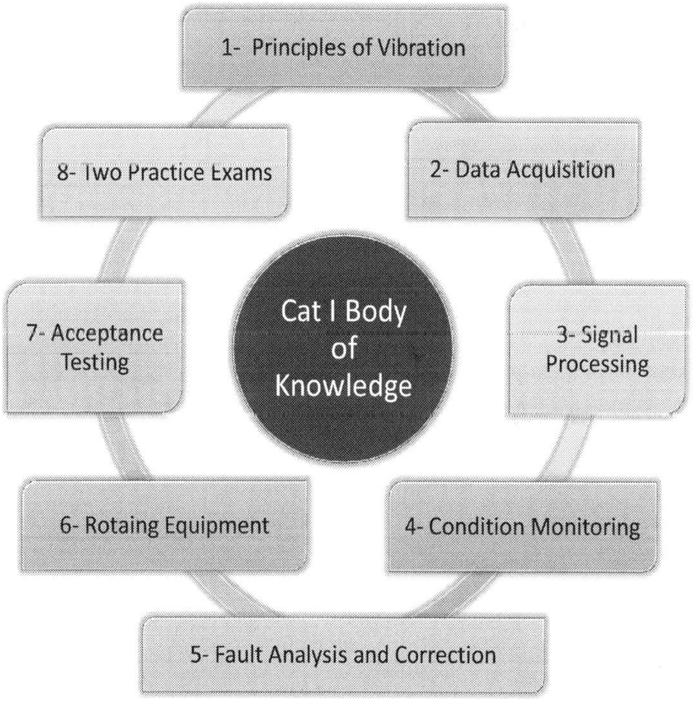

Vibration Analysis Certification Exam Preparation Package
Certified Vibration Analyst Category I
ISO 18436-2 CVA Level 1
CAT I PREP I SERIES PRACTICE TESTS

Part	Covered Body of Knowledge	ISBN-13
1	PRINCIPLES OF VIBRATION	978-1-64415-006-1
2	DATA ACQUISITION	978-1-64415-009-2
3	SIGNAL PROCESSING	978-1-64415-002-3
4	CONDITION MONITORING	978-1-64415-005-4
5	FAULT ANALYSIS AND CORRECTION	978-1-64415-008-5
6	ROTATING EQUIPMENT	978-1-64415-014-6
7	ACCEPTANCE TESTING	978-1-64415-004-7
8	TWO PRACTICE TESTS	978-1-64415-007-8

Notes

Notes

Made in United States
Orlando, FL
04 November 2026

90206053R00105